DAVID D. NOLTE

Purdue University

INTERFERENCE

The History of Optical
Interferometry and the Scientists
who Tamed Light

OXFORD
UNIVERSITY PRESS

OXFORD
UNIVERSITY PRESS

Great Clarendon Street, Oxford, OX2 6DP,
United Kingdom

Oxford University Press is a department of the University of Oxford.
It furthers the University's objective of excellence in research, scholarship,
and education by publishing worldwide. Oxford is a registered trade mark of
Oxford University Press in the UK and in certain other countries

Published in the United States of America by Oxford University Press
198 Madison Avenue, New York, NY 10016, United States of America

British Library Cataloguing in Publication Data
Data available

Library of Congress Control Number: 2023930767

ISBN 978–0–19–286976–0

DOI: 10.1093/oso/9780192869760.001.0001

Printed and bound by
CPI Group (UK) Ltd, Croydon, CR0 4YY

CONTENTS

PREFACE

The wavelength of light is the ultimate ruler. It sets the scale against which the phenomena of the universe are measured. From the imaging of distant black holes to the detection of exoplanets within our galaxy, from the generation of entangled particles in quantum computers to the transmission of photons down fiber-optic cables, from the creation of virtual reality by holographic displays to the microscopic imaging of nanoscale objects inside living cells—light helps illuminate the beauty and complexity of them all. Yet the wavelength of light is exceedingly small—about the size of a bacterial cell—a millionth of a meter. How, then, can it measure a black hole the size of our solar system that is 50 million light years away? How can it render imaginary scenes in high definition? How can it image objects that are no bigger than tiny fractions of its own length? The answer is *interference*!

Interference occurs when one wave overlaps another and they are added together, summed up, building constructively at places where peaks and troughs of the waves add together, and destructively elsewhere where peaks fill troughs and cancel out. The phenomenon of interference lies at the core of physics. Interference is why floating soap bubbles have such brilliant colors, and why they become almost invisible just before they pop. At a more profound level, interference among quantum fluctuations in the Big

Bang may be why the universe is composed almost exclusively of matter rather than antimatter. Interference is why atoms bond into molecules, why the chair you sit in is solid, and why neutron stars don't implode. Interference is also a scientific tool, allowing scientists to build gravitational wave detectors to measure the merger of black holes in distant galaxies, to launch satellites to look for alien planets orbiting nearby stars, and to build quantum computers that may one day compute incomputable things.

Interference: The History of Optical Interferometry and the Scientists who Tamed Light tells the stories behind the inventions and advancements of optical interferometry. Interferometers are the most sensitive measurement devices ever devised by mankind and are poised to change our view of our place in the universe. This book tells how superposition and interference helped shape the modern physics of optics, showing how interference and interferometers are opening new worlds that may be wilder and more alien than we imagined. The book is aimed at the curious budding scientist, at college undergraduates, as well as at experienced practitioners. It is for those who wonder where the discoveries they hear about, learn about, or use daily in their work, came from. The reader will be introduced to the personal histories of the scientists who, often working against great odds, made their discoveries, and what drove them.

Our tale begins with Thomas Young, the British polymath unusual even among polymaths, performing the first double-slit interference experiment, then turning aside to decipher Egyptian hieroglyphics. The story crosses the English Channel to France and the work of French physicists, Francoise Arago (a swashbuckling future prime minister of France) and Augustin Fresnel (banished to the countryside to build roads and bridges for

Napoleon's armies), who together fought against the entrenched forces of Laplace and Poisson to verify the wave nature of light. Arago constructed the first true interferometric apparatus and inspired a generation of French physicists during the golden age of French optics, including a fraught friendship between Léon Foucault and Hippolyte Fizeau who raced each other to use Arago's interferometer to measure the speed of light.

Our narrative then turns to the giant of interferometry—Albert Michelson—who created his iconic interferometer to measure the speed of light through the ether, only to come up empty with a null result. Disappointed with his experiments (he could not foresee that his results would help launch Einstein's relativity revolution years later), Michelson designed the first metrological interferometer to calibrate the meter bar in Paris in units of wavelengths of red cadmium light, for which he became the first US Nobel prizewinner. Michelson later extended his interferometer to astronomy where he measured the size of the star Betelgeuse in the first of a long line of increasingly sophisticated and sensitive interferometers. Descendants of Michelson's interferometer today are measuring the barest whisper of distant colliding black holes—a seeming paradox that crosses astronomical scales to detect a phenomenon a billion light years away by measuring mirror displacements that are a tiny fraction of the radius of a proton.

Moving from cosmic or atomic scales to the intermediate microscale, interference of photons creates holograms, invented by the Hungarian physicist Denis Gabor, exiled from his country by the events of WWII, while working at a British electronics company where he was isolated from the other employees because he was deemed an enemy alien. Holography captures three-dimensional information of an illuminated object by recording light interference on a two-dimensional plane to create the amazing novelty holograms we see in stores and museums. But holography goes much further, used for optical data storage and fast retrieval of massive amounts of data and for direct optical communication inside computers and in fiber-optic telecommunication systems. Holography is also used for laser ranging, much like radar, seeing into turbid media and living tissue that are opaque to ordinary optics, tracking objects moving with speeds down to nanometers per second. The underlying optical processes of holography—diffraction and interference—are the same that support optical microscopy, including surprising new methods that allow microscopes to see with finer resolution than previously thought possible.

Our story returns briefly to stellar interferometry, where the British scientists R. Hanbury Brown and Richard Twiss (known collectively as HBT) created a new type of interferometer, known as an intensity interferometer, that differed radically from Michelson's, expanding the number of stars that could be measured interferometrically. The HBT interferometer, performed on the classical light from stars, had the unexpected side effect of launching the field of quantum optics that manipulates single photons and harnesses their quantum properties to perform novel types of communication and computation. When a quantum computer calculates something much more quickly and cheaply than a classical computer, it is called "quantum advantage." It would be a calculation that takes a quantum computer minutes or hours when the fastest classical computer (which includes distributed classical computing over the cloud that can harness thousands of computational nodes) would take months or years or more. There are already reports that quantum advantage has been achieved, and more claims will be coming soon, enabled by the subtle power of optical interferometry.

These are some of the stories told here in *Interference*, gathered into a single volume for the first time, highlighting the human element in the pursuit of science. I gratefully acknowledge the help of Prof. Dan Milisavljevic at Purdue University for pre-reading the chapter on stellar interferometry and Prof. Z. H. (Jeff) Ou at the University of Hong Kong for providing suggestions for the chapter on quantum optics, and as always, I am grateful to my editor at Oxford, Sonke Adelung, for his continued support and encouragement. I also extend a thankful acknowledgment to Nicholas Nolte who read the quantum chapters and pointed out interesting improvements. This book would not have been possible without the love and support of my wife, Laura, who has always been my biggest fan.

DAVID D. NOLTE
Lafayette, IN
November 2022

1

Thomas Young Polymath

The Law of Interference

I have found so simple and so demonstrative a proof of the general law of the interference of two portions of light . . . that I think
it right to lay before the Royal Society a short statement of the
facts which appear to me so decisive.[1]

Thomas Young, *Bakerian Lecture* (1803)

The decision in 1798 by the executive committee of the French
Revolution, known as the Directory, to send an overly ambitious general to the far end of the Mediterranean Sea to conquer
Egypt like Alexander the Great, was the unlikely link among a
set of epic discoveries that changed the course of science. One of
these discoveries was the most famous linguistic decipherment in
history. Another was the most important theorem in the mathematics of functional analysis. Joining these were discoveries
that exposed the nature of light. All of these took place over the
decades following the invasion, but they were connected, like
the threads of a cobweb, to the fateful decision by the general to
comply with the Directory's wishes.

Napoleon Bonaparte, the general in question, had his eyes
on Egypt and the Middle East with a plan to sever Britain's ties
to India, but his ambitions were more than military. When he

Interference. David D. Nolte, Oxford University Press. © David D. Nolte (2023).
DOI: 10.1093/oso/9780192869760.003.0001

invaded Egypt, in addition to bringing cannons and horses and legions of soldiers—he brought a legion of *savants*. These were mathematicians and historians and botanists and philologists. These were engineers and chemists and architects and archeologists. They came with trunks packed with arcane books, delicate scientific tools, and the most advanced measuring instruments of the day. The savants, as they stepped aboard their fleet of ships sailing from Toulon near Marseilles, did not even know where they were bound, so secret had the target of the mission been kept.[2] They simply were responding to the charismatic pull of Napoleon and the promise that it would be the opportunity of a lifetime. The promise was prophetically true for those who were fortunate to return alive.

The Institute of Egypt

The French invasion landed at Alexandria and moved toward Cairo, where they were met by the Egyptian army several miles outside the city. The famous pyramids were visible on the horizon as the French routed the Egyptians—the battle later came to be known as the Battle of the Pyramids. Napoleon's army entered Cairo where officers and scientists took possession of lavish palaces abandoned by the Mameluke rulers. The newly formed Institute of Egypt set up in the palace of Hassan-Kashif on the outskirts of Cairo, still occupied by abandoned servants and an ample harem.

The president of the Institute was the mathematician Gaspard Monge (1746–1818), with Joseph Fourier (1768–1830) as the secretary and Napoleon himself as vice-president. The first meeting of the Institute was held on August 24, 1798 with 48 members. Monge and Fourier as well as Napoleon (who fancied himself good at math) were members of the mathematics section. The physics and natural history section included Claude Berthollet (a chemist who established the principles of chemical equilibria), René-Nicolas Desgenettes (a doctor and close friend of Napoleon), Déodat de Dolomieu (the rock type known as dolomite is named after him), and Geoffroy Saint-Hilaire (a colleague and proponent of Lamark). The literature and arts section included Baron Dominique Denon (the first director of the Louvre Museum). Although the mission of the Institute was ostensibly to develop the fledgling field of Egyptology, Napoleon valued it as an intelligence agency in an enemy land.

In time, this grand adventure that had seduced the savants with the desire to unlock the mysteries of Egypt turned from a dream into a nightmare. The British fleet under Horatio Nelson engaged the French during the Battle of the Nile in Aboukir Bay that raged for three days from August 1 to August 3, 1798. The French ships were destroyed, captured, or scattered. Direct contact and resupply from France were cut off, and the prospects for returning home dimmed. Despite the defeat, Napoleon did not abandon his campaign, and his troops advanced into Syria, passing through Gaza to besiege the Turkish garrison in the port city of Jaffa in present-day Israel.

Among the military engineers who accompanied the campaign was Étienne Louis Malus (1775–1812), a member of the Institute of Egypt in the mathematics branch. Malus was the son of a financial advisor to Louis XVI. At the outbreak of the French Revolution he had to leave school to join the army to avoid scrutiny, prison, or worse because of his association with the old regime.

His commanders, surprisingly cognizant, recognized his strong mathematical talent and transferred him in 1794 to the Ecole Polytechnique in Paris where he was among the first students to attend the new school. While at the Ecole, he came to the attention of its director Monge, who took a personal interest in him. After graduation, Malus returned to the army as an engineer, and when Napoleon asked Monge to assemble a corps of savants to accompany the Egyptian campaign, Malus was one of the first he picked.

The siege of Jaffa was over quickly, but the aftermath was endless. Malus wrote, "The tumult and the carnage, the broken doors, the homes shaken by the noise of the fighting and of arms, the screaming of the women … furious soldiers responding to the cries of the wounded by cries of rage and repeated blows, finally of men satisfied of blood and gold, falling of weariness on the heaps of cadavers … ."[3] Napoleon was merciless to the Turkish prisoners. Four thousand Turks surrendered and were bayonetted to death on the beach because Napoleon did not want to waste bullets. Another evil outcome of the sacking of Jaffa was the pestilence it unleashed on the French victors, and Malus recognized the event: "The frenzied pillage delivered a miasma. It was contained in the clothes they had greedily taken. The mortal effect was rapid. The illness appeared on the battlefield, in buboes and carbuncles. The horrifying cry 'It's the plague!' spread through the army and struck terror in the most courageous and invincible."[4]

As a military engineer on the campaign, Malus was ordered to set up a hospital to hold the rapidly increasing number of plague victims. He took over a convent to outfit it as a hospital, but it was understaffed and overwhelmed by numbers. When Napoleon marched up the coast to attack Acre, those who got the plague were shipped back to Jaffa where soldiers were dying at the rate of 30 per day, continuing for six weeks. Within ten days of setting up the hospital, Malus was stricken with the disease and fell into a nightmarish sequence of fevers and confusion. He was near death when he wrote letters in a brief moment of clarity saying goodbye to his friends and family. By luck, a transport ship arrived to relieve the press of patients, and it transported many back to Cairo, Malus among them. As soon as he breathed the fresh sea air he began to recover and was finally able to eat by the time he arrived in Cairo. Unfortunately, all patients were quarantined in a prison where the only medical staff consisted of the grave diggers who gave no care to the patients and stripped them of their possessions before they were even dead. Miraculously, Malus survived this ordeal, but the French could not survive in Egypt.

Toward the end of 1801, as the French resources finally gave out, they had nothing left to defend themselves from the British. Those of the savants who remained (only two thirds of the original group were left) hoped to be allowed safe passage home, but the French commander General Menou would not let them go. When the French finally surrendered, Menou cared little for the tremendous cache of notes and drawings and artifacts that the savants had amassed during their time in Egypt, and he used it as a bargaining chip in the terms for surrender, promising to give it all to the British in return for other concessions. The savants were appalled. When the British moved to confiscate this treasure trove, Geoffroy Sainte-Hilaire delivered an impassioned speech that made him famous in all the French societies.

You are taking from us our collections, our drawings, our plans, our copies of hieroglyphics, but who will give you the key to all of this? They are only preliminary drafts that our personal impressions, our observations, our memories must complete. Without us, these materials are a dead language, from which you will hear nothing, neither you nor your savants. We have spent three years conquering, one by one, these riches, three years gathering them from all of the corners of Egypt, from Philae to Rosetta: to each of them is attached a peril surmounted, a monument seen and engraved in our memories.[5]

He then stated with conviction that if the British moved to take the materials, the French would destroy it all before they had the chance. He asked what the world would think if a treasure as great as the Alexandrian Library were burned once again? The British relented, letting the savants keep their materials—except for one artifact that the British coveted more than all their other manuscripts and relics.

Passage home for the survivors was taken on a motley array of British and Greek vessels, some mere trawlers, suffering dangers as great as those faced in Egypt. When they limped into Marseilles, the savants were sequestered into quarantine for a month. Only in 1802 did they finally walk free onto French soil. Later, as Napoleon rose to take absolute power, he rewarded his most faithful servants, making Berthollet and Monge counts. Berthollet set up a society in Arcueil, where a recovered Malus became a member, giving him the freedom to pursue ideas that he had conceived in Egypt on the properties of light. Laplace also joined the small group, and he took Malus under his wing.

Sainte-Hilaire's words were prophetic; the vast collection of diverse materials the savants carried back to France became the source for *Description de l'Egypte*, a comprehensive description of the ancient world that the savants had rediscovered during their expedition. The task of preparing the work for publication was taken on by Fourier. He had recognized the unique opportunity as early as 1798 to shed light on the mysterious land on the far side of their own Mediterranean Sea that had captured the imagination of scholars for centuries. The first volume was edited and published by Fourier in 1809 under the direction of Napoleon (always looking for an opportunity for self-aggrandizement). Many more volumes appeared later under the auspices of the restored King Louis XVIII after 1814, with the final volume published in 1829. The effort took 30 years, consisting of 24 volumes, compiling the work of the 150 savants who sailed with Napoleon from Toulon in 1798 without knowing their destination. The *Description* drew from the vast materials that the British had obligingly returned, but they retained that one most coveted item—the Rosetta Stone.

The Linguist and the Loyalist

A year after the Battle of the Pyramids, when the French were still digging in, a gang of soldiers uncovered a nearly one-ton shattered slab of black basalt in the Nile delta near the town of Rashid, called Rosette by the French and Rosetta by the British. Its one polished surface was inscribed with three types of writing: one was Greek, used to write the Coptic language of Egypt; another was demotic, the everyday script of later Egyptian antiquity; and the last was the long-forgotten script of Egyptian hieroglyphics. The inscription was a commemoration of Ptolemy V Epiphanes, dated to 196 BC, understood through the Coptic version of the text, which scholars already could decipher. The trilingual inscriptions were assumed to be equivalent translations, but while the demotic portion was almost entirely intact, the Coptic was missing several parts, and the hieroglyphic section was badly damaged and not even half complete. Nonetheless, Joseph Fourier and Étienne-Louis Malus inspected the stone and suspected that the Rosetta Stone could be the key to decipherment of Egyptian hieroglyphics.

At the time of the French invasion, Egyptian hieroglyphics were mistakenly believed to be a primal visual language that communicated wisdom directly through symbols imbued with mystical meaning. Even in the seventeenth century, mathematician and philosopher Willhelm Gottfried Leibniz (1646–1716) was searching for a universal "characteristic"—a language of pure logic based on arrangements of mathematical symbols, and he viewed hieroglyphics as an archetype. For these reasons, the decipherment of Egyptian hieroglyphics presented one of the great opportunities of the Enlightenment. By the single stroke of a pickaxe, the key to eventual decipherment had been uncovered. Within months of the discovery, despite the waging war, French scientists made copies of the inscriptions and sent them to their colleagues across Europe. A decade passed with little progress until the British physician and physicist Thomas Young (1773–1829) turned his penetrating mind to the problem of the decipherment.

At the beginning of 1814, Young received an envelope in the post that contained a copy of a damaged papyrus manuscript containing demotic script. At that time, he was a 40-year-old London doctor whose medical practice suffered for want of patients as his attention tended to wander over many non-medical subjects such as linguistics and orthography. He had become known to have a great facility with writing systems, able to recognize and to reproduce original texts with meticulous care. This was one of his chief talents—the ability to see simple patterns and to extract the salient details from a vast visual jumble of information. As he worked on the papyrus, he naturally began to think about the Rosetta Stone, which had found its way to the British Museum in London, stubbornly thwarting the efforts of professional linguists. To augment his understanding of Egyptian writing, Young turned to the published volumes of Fourier's *Description de l'Égypte*.[6] The book contained a treasure trove of exquisitely rendered colored plates of inscriptions and papyri. By the middle of the year, Young was immersed in Egyptian studies, assimilating all he could find, comparing and contrasting scripts and figures and characters, noting all the similarities and all the variations. In the midst of this activity, he received another letter in the post from a friend who knew of Young's intense interests.

Young must have thrilled with excitement as he opened it, finding the reproduction of a recently uncovered funerary papyrus of ancient Egypt. The papyrus carried two scripts—one the running demotic script, and the other the hieratic script, which looked vaguely similar to hieroglyphics, but written with ink on papyrus instead of carved in stone. As he scanned the new text, Young must have been struck by an apparent similarity between several of the demotic characters and their hieratic counterparts. Turning to the reproduction of the Rosetta Stone, he isolated several of the same demotic characters, and then compared them against the hieroglyphic, finding the same striking similarities. In a flash of insight, he realized that demotic must be, at least in part, a modified version of hieroglyphic writing, somewhat like cursive writing is related to printed writing today.

To confirm his hunch, Young turned to the groups of characters in the hieroglyphic portions of the stone text contained in
a *cartouche*. A cartouche is a group of characters separated from
the rest of the text, enclosed by an oblong oval that resembled
powder cartridges, known as cartouches, carried by the French
soldiers of the Egyptian Expedition. The Coptic version of the
Rosetta text referred to Ptolemy in several places, and Young
located the single surviving cartouche in the hieroglyphic section
that was likely to carry that name. The correspondence was not
one-to-one, but Young was able to isolate several symbols in
common between the demotic and the hieroglyphic. By looking
for further correspondences among the texts (see Figure. 1.1 for a
comparison), Young correctly identified the writing direction as
right-to-left. Because demotic was largely a phonetic script, with
characters representing sounds, Young had discovered that many
Egyptian hieroglyphs were phonographic in nature, shattering
the romantic notion of the magic symbols.

Hieroglyphs

Hieratic

Demotic

Coptic

Figure 1.1 Hieroglyphs
compared to hieratic,
demotic, and Coptic
scripts. The Coptic
script was a form of
Greek.

These discoveries by Young—that many demotic characters were a "cursive" form of hieroglyphic, and that many hieroglyphic characters were phonetic—allowed him to create translations of several proper names in the hieroglyphic text, and to identify the phonetic values of several of the hieroglyphic characters. This was the greatest inroad yet made into deciphering Egyptian hieroglyphics. Young became semi-famous for this feat, his name recognized in wide circles. He was asked to write the 1818 entry on Egypt for the *Encyclopedia Britannica*, an article that contained many of his insights and discoveries in the field.

His success in the field of Egyptology was the latest of a string of successes that marked his singularly peripatetic career. He was, after all, merely a city doctor who took summers off from his practice to pursue a host of unrelated activities, such as directing the Bureau of Longitude as an expert in navigation, and consulting for the British Admiralty as an expert on the construction of ships. Yet all of these significant activities paled in relation to the work he had accomplished at the very beginning of his career. Long before Young became a practicing doctor with too many sidelines, he was one of England's leading physicists, his interest in the subject arising from his medical studies of the eye and ear which had led him to study the physics of light and sound.

Young's upbringing had been strict but not poor, born to Quaker parents in the west of England near the Bristol Channel that snakes its way out to the Celtic Sea between Cornwall and Wales. Although his schooling was typical of the period, he was fundamentally an autodidact, teaching himself to read by the age of two and completing a thorough reading of the Bible by the age of six. As a schoolboy, he immersed himself in studies of natural history and natural philosophy as well as languages, reading

Newton's *Principia* and *Opticks*, Lavoisier's *Elements of Chemistry*, Joseph Black's lectures on chemistry, and Boerhaave's *Methoclus studii medici*.[7] He independently studied Hebrew, Chaldean, Syriac, Samaritan, Arabic, Persian, Turkish, and Ethiopic in addition to the usual languages like Latin and Greek.

Young's first publication was in 1793 when he was 20 years old in his first year of medical school at the University in London. Part of the instruction involved the routine dissection of the eye of an ox. Young turned the routine exercise into a ground-breaking medical experiment in which he identified the mechanism by which the eye focuses at different distances (a process called accommodation) by isolating the key role of the crystalline lens and the muscles that change its curvature. As part of Young's studies, he painfully poked thin keys into his own eyes to measure their curvatures and diameters. He also discovered the origin of astigmatism, caused by the asymmetrical shape of some corneas. The paper he published in the *Philosophical Transactions* of the Royal Society was so well received that he was elected a member of the Royal Society the next year.

Despite his talents and success, he was a Quaker, which was not a religion sanctioned by the state, preventing him from earning a medical degree from an English university. Unwilling just yet to give up his family's religion, he enrolled in the medical school at Edinburgh in Scotland, transferring one year later to the medical school at the University in Göttingen in Germany where he received his MD in 1796. When he returned from Germany in 1797, Young was still required to obtain a degree at an English university before being allowed to practice medicine in England. By this time his dedication to his Quaker upbringing had conveniently faded, so he converted to the Church of England and entered Emmanuel

College at Cambridge where he made contact with many top names in learned society, including Edmund Burke and William Herschel.

Young's studies at Cambridge reignited his earlier interests in physiological perception, especially regarding how the eye sees and the ear hears. Either through his genius for making connections among seemingly disparate topics, or through his inability to keep his attention on one subject, his physiological studies led him into physical acoustics and the sounds of organ pipes. He investigated the relationship between the length of the pipe and the wavelength of the standing pressure wave, paying particular attention to the role played by internal reflections in establishing the resonance in the tube. As his wide-ranging mind looked at the pattern of rows of organ pipes, arrayed in increasing lengths high on the walls of a church, he may have remembered seeing a similar pattern, on an unrelated subject, from his early days reading Newton's *Optiks* on the subject of colored films. When he reread the *Optiks* he must have been struck by the strong similarities between the periodic patterns of colors of light in thin films (Newton's rings) and the periodic patterns of organ pipes resonating with sound.

This was not the first time someone with insight had seen this similarity. Leonhard Euler (1707–1783) had made this analogy nearly 50 years before,[8] thinking of the colors in thin sheets of glass like vibrating strings composed of oscillating particles of ether. It took only simple arithmetic to relate the thickness of the film to the wavelength of the light, and Euler had proposed that the different colors of light arose from different wavelengths of the vibrating ether.[9] However, Euler's opinion was a minority view at the time,[10] as Newton's emissionist theory of particles of light held sway over nearly all natural philosophers. In New-ton's theory, the colors of films were attributed to periodic "fits" in the material—the emission theory admitted no vibratory explanation.

Young saw it differently. As he came to understand the partial reflection of sound waves from the open end of an organ pipe that set up a standing wave inside, it made him look at the thin glass sheet and think of partial reflections of light at the interface between the air and the transparent medium. In this view, the thin sheet was not resonating in the way Euler had thought, as a vibrating string where the physical material was oscillating, rather the partial reflection of light from each interface caused a standing wave of light to occur inside the thin film, just like the standing waves of sound in the organ pipes. Young must have been captivated by this close analogy, in particular with this insight into the important role of partial reflection, and he wrote up a detailed "Outline" that documented his ideas, including cal-culations on the relationship between color and wavelength. This was a defining moment for Young, when he committed himself to the wave nature of light despite the overwhelming majority opinion against it.

After graduating from Cambridge in 1800, Young moved to London with the hopes of establishing a medical practice, having inherited the house of a rich uncle. Unfortunately, Young was socially awkward, and his medical practice never succeeded to the same degree that his other endeavors had. He was at heart a gentleman scholar, and the inheritance from his uncle, if not immense, was at least large enough to allow Young to follow his many interests comfortably. Once in London, he attended meetings of the Royal Society and soon was acclaimed as one of the leading figures in the physiology of perception. He read a paper "Sound and Light" to the Royal Society in 1800 and was selected as that year's winner of the Society's Bakerian Lecture,[11] delivering his speech "On the Mechanism of the Eye."

Buoyed by the reception of his researches into the physiological aspects of light, Young embarked on a more focused treatment of the partial reflection of light and the generation of standing waves. He envisioned light as periodic patterns that moved back and forth between interfaces, adding together in a complicated way that caused bright and dim regions inside the thin film, like the formation of moiré patterns seen by overlaying two fine cloths. He was still groping in the dark on the actual nature of the light waves, not fully understanding how the partial waves would superpose, but this was a glimpse of a first crude principle of interference.

Young was selected again for the Bakerian Lecture of 1801, this time giving his lecture "On the Theory of Light and Colours." Young had expanded his earlier physiological studies into color perception in the eye and proposed that the eye saw the rainbow of perceived colors using only three color receptors: red, green, and violet. For instance, yellow could be perceived as the mixture of red and green light, and blue as the mixture of green and violet. His grasp of the tricolor theory of color perception was helped in part by his acquaintance with John Dalton, who was blind to the color red and who described a perception of colors that differed from what most others saw. Young performed simple demonstrations that left little doubt about the general validity of color additivity. He then took a step further to explain that the color receptors in the eye vibrated in resonance when driven by the different frequencies of light. Using estimates of the thickness of films and colored fringes, Young estimated the wavelength of light, which, when combined with the speed of light, yielded a temporal frequency that caused the receptors in the eye to vibrate.

Young's second Bakerian Lecture in November of 1801 was pivotal for two reasons. First, he openly supported the wave nature of light and proposed a first crude principle of interference. This aspect of his lecture was considered intriguing but too speculative to be accepted. It also occasioned the first pointed attacks on Young by others who firmly held to the emission theory of light. Second, it established Young as the leading scientist in Britain in the branch of natural philosophy concerned with physiological perception and more generally with the physics of sound and light. This put him in an excellent position to be considered for a position as lecturer at the Royal Institution, newly founded by the ex-American Benjamin Thompson, a Royal Loyalist who spied against the Americans in the Revolution, escaped to London, and later became a Bavarian count known as Count Rumford.

Benjamin Thompson was born on a poor farm in a small town near Boston. Despite no formal schooling, he was intelligent and an autodidact, like Young, and never passed up an opportunity for advancement. In the years preceding the Declaration of Independence, Thompson joined a British Loyalist brigade and helped spy on American "agitators" to his British superiors. However, after the American victory at Bunker Hill he was forced to evacuate with Howe's troops from Boston, fleeing to England, where he was appointed to successively higher positions of authority over colonial affairs until the end of the war.[12] Always restless and looking for opportunities, he was recruited as a military advisor to Bavaria for 16 years, restructuring its military, improving public works, and acquiring a peerage with the title of Count. Because he was a man without a country, with a flair for the ironic, he took the name of Rumford, the older name of Concord, New Hampshire, where he had lived before the war.

As Count Rumford was overseeing the boring of cannon in Munich, he noticed how hot the cannon barrels became, caused by the friction from the bore, in some cases becoming red hot. The incessant motion of the bore was converted into heat, contradicting the prevailing theory that heat was a substance, and Rumford realized that heat must be a form of motion. He wrote up his treatise "An Experimental Enquiry Concerning the Source of the Heat which is Excited by Friction"[13] published in the *Philosophical Transactions* in 1798. It was one of the first serious attacks against the caloric (particulate) theory of heat. This publication is hailed as a seminal step in the direction of the conservation of energy, which would develop over the next 50 years.

Returning to England shortly before Young moved to London, Rumford helped found the Royal Institution in 1799 with a royal charter to advance practical applications of science and to serve as a conduit to disseminate the new knowledge to the wider public. Public lectures became the cornerstone of the Royal Institution, and he hired Humphrey Davy as the first lecturer, who subsequently hired Michael Faraday to be his assistant. In addition to popular lectures, the Institution also provided more specialized seminars. In 1801, Joseph Banks, the president of the Royal Society, was impressed by the breadth of Young's knowledge, especially his mastery of all branches of natural philosophy, and Banks suggested Young as a candidate to Rumford as a lecturer in natural philosophy. Young made a favorable impression on Rumford when they met, and on August 3, 1801 Young was selected as "Professor of Natural Philosophy, Editor of the Journals, and

Superintendent of the House" at the Royal Institution with an annual salary of £300. He was to prepare a full set of lectures on the state of natural philosophy to begin in January 1802.

Young had been gaining momentum in his scientific activities through the end of his Cambridge days and into his first years in London, yet the character of his research had remained superficial, consisting of astute speculations and the broad analogies of a dilettante. So, when he was hired by Rumford to lecture on the state of natural philosophy at the turn of the new century, he took advantage of the opportunity to become a *bona fide* natural philosopher. He threw himself into the task as only someone with the supreme abilities of an autodidact can do, reading through the existing literature, old and new. His goal was to collect and synthesize all that was known into a coherent overview that captured the current state of the mechanical arts. During this phase of his self-education, one of the topics that he familiarized himself with was the mathematical theories of Christiaan Huygens on the wave nature of light.

Christiaan Huygens' Principle

Christiaan Huygens (1629–1695) was a seventeenth-century Dutch mathematician and physicist, known today primarily for the invention of the pendulum clock and for expounding what came to be known as Huygens' principle for the propagation of light. His *invention* of the clock dealt with the mechanics of ponderable masses and centrifugal forces, while his *principle* dealt with ethereal light and the imponderable passage of waves through matter. While *invention* and *principle* seem to have no common physical basis, both were born from the same mathematical tools that Huygens developed to explain the physical world. In one of the first examples of mathematical physics, Huygens unified the descriptions of clocks and refracted light.

Christiaan Huygens was born in April 1629 in The Hague, the *de facto* capital of the United Provinces of the Netherlands, or the Dutch Republic, that had won its freedom from Spanish rule only a few decades previously. His father was an advisor to the House of Orange and moved in high circles of elite diplomats, famous artists, and enlightened scientists, providing Christiaan with an unsurpassed educational environment. Although the father hoped this second son would enter law, mathematics was a stronger calling, and Christiaan learned from the master mathematician Frans van Schooten (1616–1660), who had translated and written a commentary on Descartes' *Géométrie*. Descartes introduced analytic geometry to the world, but it was van Schooten who introduced Descartes to the world. Under van Schooten's tutelage, Huygens made original mathematical contributions of his own. He corresponded with his father's friend Marin Mersenne (1588–1648) in France who was so impressed by the young Huygens that he christened him the "new Archimedes."[14]

Huygens' academic years were highly productive. He became an early master of analytical geometry and applied it to longstanding problems in conic sections. When he returned home from the university in 1654, he became interested in telescopes and the optics of lenses, and he set up his own observatory. The superior quality of his telescope lenses that he ground himself allowed him to discover Titan, the largest moon of Saturn. He also was able to discern for the first time that the strange "appendages" of Saturn, discovered earlier by Galileo and thought to be attached to the planet, were in fact thin rings hovering disconnected from the planet's body. His description of these discoveries, transmitted through his many letters to friends and scientists, made him famous across Europe.

A year later, Huygens turned his attention to clocks. Well-regulated clocks were key astronomical instruments for keeping track of the minute motions of celestial objects. He was also motivated by the goal of finding an accurate way of keeping time to aid in the navigation of ships by determining their longitude while at sea. Huygens was aware of Galileo's work on the physics of the pendulum and its potential for time keeping. The pendulum has a natural frequency that is mostly independent of the amplitude of its swing, especially for small amplitudes. But Galileo had not been able to devise an effective mechanism to keep the pendulum from winding down due to friction. Sometime late in 1656, Huygens invented a way to couple a swinging pendulum to the verge escapement mechanism that had been used in the old "verge and foliot" clocks for over a century. Huygens' verge escapement had a crown gear with asymmetric teeth which, when held under tension by hanging weights or springs, gave a little "tap" to the pendulum on each cycle, thereby keeping it swinging as long as the tension was maintained. He engaged a clock maker in The Hague to build a working model, and he promptly applied for a patent in 1657 and began looking for ways to improve it.

By Huygens' time it was known that the pendulum was not strictly isochronous but would swing more slowly as its angular displacement became larger. Unfortunately, Huygens' verge escapement required large swings for the verge to clear the teeth of the crown gear, and so it suffered from this non-isochronicity. To address this problem, Huygens began the search for the tautochrone curve—a curve along which a mass would oscillate with constant period independent of its amplitude. His mathematical explorations took him down untrodden paths of curves and tangents and envelopes that he addressed with techniques of analytical geometry. (The invention of the calculus was still a few years away.) What Huygens developed during his search for the tautochrone was a theory of evolutes and involutes.[15]

The evolute of a curve is the set of all of its centers of curvature. For a continuous curve, like a parabola or an ellipse, the evolute is also a curve. Part of what Huygens accomplished was to show that the evolute is also the envelope to all of the normals of a curve. This is seen in Figure. 1.2, showing the envelope of the normals to an ellipse. Closely related to the evolute is the involute. The involute of a curve is what would be traced out by the tip of a taut string that is initially wrapped onto the curve and is then "unwound" until it is straight. This is also known as the rectification of a curve, where one finds the equivalent straight line that has the same length as the arc of the curve. Huygens was

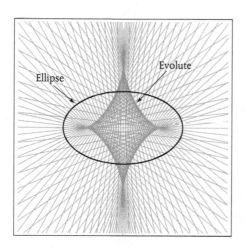

Figure 1.2 Evolute of an ellipse constructed by drawing many normals to the ellipse. The envelope of these normals, which emerges automatically in this construction, defines the evolute. The evolute is also the locus of all the centers of curvature.

particularly interested in involutes, because he reasoned that his clock could become isochronous if he added "bumpers" around which the string of the pendulum would partially wind itself. Then the curve traced out by the pendulum bob would be exactly the involute of the shape of the bumper.

Huygens proved that the involute of an evolute of a curve is the curve itself. Therefore, if the tautochrone is the involute of the bumper shape, then the evolute of the tautochrone is also the bumper shape. If this all sounds a bit circuitous, it is, because Huygens finally showed that the curve that is its own involute, and hence is its own evolute, is the cycloid, leading to the discovery of the tautochrone in 1659. If the bumper on his pendulum clock took the shape of a cycloid, as in Figure. 1.3, then the arc path of the pendulum bob also would be a cycloid

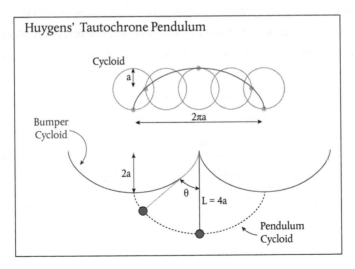

Figure 1.3 Huygens' bumper to make a pendulum clock isochronous. The tautochrone curve (the pendulum cycloid) is the involute of the bumper.

and would be isochronous for all amplitudes of motions up until the pendulum string was wrapped entirely around the bumper. Once Huygens knew what shape to make the bumper, he had prototypes made and tested. The tests went well enough for Huygens to know he had been right, but factors such as the additional friction of the string against the bumper and details of the escapement mechanism still limited the accuracy of his clocks. But the most important result that came from Huygens' mathematical explorations was this new mathematical tool in his toolbox—this idea of families of curves—which would serve him decisively in the entirely different field of optics, as we shall see shortly.

The pendulum clock was a disruptive technology of its day. Its accuracy far outstripped the older technology, and pendulum clocks quickly swept across Europe. Patent law at that time was not as firmly established as today, and bootleg copies of Huygens' clocks became much more common than his own, robbing him of significant income. Furthermore, around 1658 Robert Hooke devised a much better escapement than Huygens' verge. The new escapement is known as the anchor escapement and it allowed for longer pendulums with smaller swings (hence the origin of so-called grandfather clocks) that had better accuracy. But what Huygens lost in monetary value was compensated by what he gained in reputation, and in 1665 he was invited to become a foreign member of the prestigious Royal Society of London. One year later, King Louis XIV of France, at the recommendation of his first minister Jean-Baptiste Colbert, inaugurated his own *Académie des Sciences* in Paris, and Colbert invited Huygens to fill a new position at the *Académie*. Huygens accepted and travelled to France in 1666.

His new life in Paris, and his new responsibilities, were both a blessing and a burden. He was exposed to elevated minds at the top of their fields who also had ruthless ambitions and unbridled jealousies. He was also exposed to a tense social life that was strictly formalized and overly aggressive for his delicate constitution. Although Huygens had been recruited because of his reputation as an astronomer through his discoveries associated with the planet Saturn, he followed his own interests, creating friction with the *Académie* that would grow across the years to an eventual breaking point. Even at the start, Huygens must have sensed the mismatch of his delicate character to the rough-and-tumble society he had entered, and it threw him into one of his first bouts of melancholy, forcing him to return home temporarily to The Hague in 1670 to recuperate. His absence at just this moment from the *Académie* was crucial, because he otherwise would probably have been selected for an important astronomical expedition that changed how we see our place in the cosmos.

In 1671 the French *Académie* sent an expedition led by Jean Picard (1620–1682) to the former location of Tycho Brahe's Uraniborg observatory on the island of Hven outside Copenhagen. Tycho had abandoned the site in 1596 when he was driven out by a new king suspicious of his strange science, which looked like witchcraft. The goal of the French expedition was to test a proposal made by Galileo 50 years earlier to the King of Spain that longitude could be determined by timing the eclipses of the moons (the Galilean moons) of Jupiter, using them like the hands of a precisely timed clock that could not be swayed by a heaving deck at sea. Arriving in Denmark, Picard engaged the services of the Danish astronomer Ole Rømer (1644–1710) to observe more than a hundred eclipses of the Galilean moon Io by Jupiter.

The same eclipses were being observed by Giovanni Domenico Cassini (1625–1712) in Paris as both teams kept precise records of the times of the eclipses. The timings were expected to differ by an amount related to the difference in longitude between Paris and Hven. Although this explicit goal did not succeed, the Uraniborg expedition was fortunate to have two unintended consequences for the history of science. One was the observation of a systematic shift in the timing of the eclipses that depended on the time of year, leading to the later discovery of the finite speed of light by Rømer. The other consequence derived from the large clear crystals of Iceland spar that Picard brought back to Paris.

Iceland spar is a transparent form of calcium carbonate, calcite, that had been mined for centuries as large crystals from a geothermal vent at Helgustathir on the eastern coast of Iceland.[16] The Vikings are thought to have used the crystals as "sun stones" to help them find the location of the sun on cloudy days to navigate. The crystals take in natural light and split it into two polarizations with a slight spatial shift that allows them to be used as a polarizer, like a natural type of Polaroid sunglasses. In the Arctic, light scattered from small ice crystals high in the atmosphere takes on partial polarization, depending on the direction of the sun, which could be analyzed using the crystals. In 1669, some of these crystals found their way into the hands of Rasmus Bartholin (1625–1698) at the University of Denmark, who carefully studied the double image that was formed when looking through the crystal. He realized that the famous Snell's law of refraction was violated by one of the images. In particular, a beam of light entering the crystal perpendicular to a crystal face (normal incidence) refracted into a small angle. Snell's law of sines forbade any such deflection of the light. Bartholin published his findings on this extraordinary behavior of light, but no explanation was forthcoming. When Picard was about to return to Paris, Bartholin gave him a large crystal of Iceland spar. At that time Rømer was Bartholin's student and was living in his house (he would eventually become his son-in-law), and when Rømer followed Picard to Paris a year later, he brought more crystals with him. In Paris, the crystals began to circulate among the members of the Académie, one of whom was Huygens who had recently recuperated and returned from The Hague.

1672 was a watershed year for Huygens. Not only was it the year when Iceland spar was brought to his attention, but it was the year that he received a visit from a bright young lawyer representing the German state of Hanover with an audacious proposal for King Louis XIV, hoping to convince the war-minded French monarch of the merits of a military invasion of Egypt—not unlike the plan that sent Napoleon to Egypt over a hundred years later. This German rube was never allowed to see the King nor even his first minister, but he did somehow wrangle a meeting with Huygens because of his side interest in mathematics. Huygens was impressed with the visitor and agreed to tutor him in the latest methods of analytical geometry. The upstart German lawyer was Gottfried Leibniz. Huygens could not have realized that he would later be surpassed by this student who would go on to invent the differential and integral calculus. Despite outward appearances, Leibniz' plan for a French invasion of Egypt was not capricious—it was an attempt to divert an impending war whose consequences would ripple across Europe, embroiling his own state of Hanover as well as Huygens'.

In 1672 Louis XIV launched a massive military campaign to conquer the United Provinces of the Netherlands. France was supported briefly by England, while the Holy Roman Empire (including the Electorate of Hanover) and Brandenburg-Prussia supported the Dutch. The initial assault was rapid as the French troops swept across the Spanish Netherlands into the Dutch Republic, nearly to Amsterdam itself when the Dutch force, under orders from the stadtholder William III of Orange (and future King of England), breached the dykes and flooded the polders that lay before the French army, stopping their advance. The war bogged down for six long years before the French forces finally were beaten back.

During this action, Christiaan Huygens' older brother, Constantijn Huygens, served as William's personal secretary—the Huygens family, after all, were longstanding servants of the House of Orange. Yet Christiaan Huygens served the French King who had unleashed this onslaught. Oddly, Huygens' activities were not affected much by the war, and when he published his triumphant analysis of oscillatory motion and clocks in 1673 in *Horologium Oscillatorium (The Pendulum Clock: or Geometrical Demonstrations Concerning the Motion of Pendula as Applied to Clocks)*, he included an overtly obsequious preface praising Louis XIV. Why he did this, without even a hint of irony, has puzzled historians for centuries.[17] Yet, even at the height of the war, with his brother in its midst at the side of King Louis' nemesis, Huygens was able to find ways to skirt the front lines to visit his father.[18]

When Huygens first got his hands on one of the crystals of Iceland spar, he was captivated, and a little disturbed, by the extraordinary refraction that violated Snell's law. Because of his interest in the properties of lenses and imaging in telescopes, Huygens considered himself to be an expert on the science of refraction, a topic known as dioptrics, and he had been working on and off for years on a treatise on dioptrics. At the heart of dioptrics was Snell's law—a (formerly) inviolate law of physics that governed all that anyone could know about the action of lenses and their role in imaging. A study in dioptrics always started with Snell's law, from which one derived its consequences on image quality. Yet here, in Iceland spar, the law was violated, violating everything that Huygens had assumed about the behavior of light, and he resolved to explain it.

In the midst of this intellectual crisis in dioptrics, Huygens read a series of letters published in the *Transactions of the Royal Society of London* of 1672 between a young mathematician in England, Isaac Newton (1642–1727), and a French Jesuit priest, Ignace-Gaston Pardies (1636–1673), on the nature of light. Pardies was critical of Newton's experiments on dispersion of colors in prisms and took the opportunity to lay out a wave theory of light that contained three essential assumptions: that light did not travel instantaneously but took finite time, that it traveled in the form of a wavefront, and that the directions of rays were perpendicular to the wavefront. Newton's response, based on the corpuscular theory of light, was forceful enough that Pardies recanted, but not before he had personally shown Huygens his theory. The idea lodged in Huygens' mind and stayed there, no matter how forceful Newton's arguments were. But despite Huygens' efforts, the extraordinary refraction in Iceland spar would not yield to analysis, and he was forced to lay his dioptrics aside until some new approach presented itself.

Beginning in 1672 Ole Rømer became Cassini's assistant in Paris, and the two continued making accurate timings of the eclipses of Jupiter's moons to complement the measurements made in Hven with Picard. Over the following years at the Paris

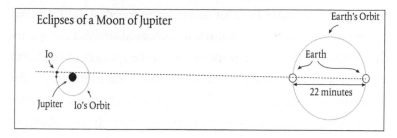

Figure 1.4 Assuming that Io's orbit around Jupiter is like a perfect clock, the exact time when its eclipse by Jupiter is observed on Earth varies by 22 minutes depending on whether the Earth is closer or farther from Jupiter. It takes light 22 minutes to traverse the diameter of the Earth's orbit.

observatory, Rømer discovered a yearly trend in the observed times of the eclipses with a surprisingly large shift of approximately 22 minutes over the year. In a flash of insight, he guessed that the effect was caused by the finite time it took light to traverse the diameter of the orbit of Earth around the sun, shown in Figure. 1.4. Cassini initially was receptive to the idea that the delay was caused by the finite speed of light and even proposed it in a letter he read before the *Académie* in August 1676. Yet Cassini wavered as he considered other explanations. Galileo had shown that the speed of light was exceedingly fast—too fast to measure using terrestrial means, and Cassini was not bold enough to refute the master.

Rømer, on the other hand, was convinced that the effect was caused by the finite time it took light to travel the longer or shorter distances to Jupiter as the Earth changed position around the sun. After he reviewed the measurements that he had made with Picard in 1671 and the observations made with Cassini, he presented his own paper before the *Académie* in December 1676, making an argument in favor of the finite speed of light. His reasoning was bolstered by careful mathematical analysis comparing the motion of the Galilean moons with the motion of the Earth, constrained by nearly eight years of careful measurements. In this way, Rømer measured something that was supposedly too fast to measure, and he measured it using the leisurely and deliberate motions of distant objects in the solar system. Rømer was not in possession of all the unknowns he needed to solve for an actual value for the speed because he was not an expert in light, but then there was Huygens.

It says something about Huygens' dual relationships with the *Royal Society of London* and the *Académie des Sciences* in Paris that he was not present or even aware of the discussions of the finite speed of light at the *Académie*, but instead read about it in the June 25, 1677 issue of the *Transactions of the Royal Society*.[19] Huygens' reaction was immediate. By considering the orbital radius of the Earth in relationship to the orbit of Jupiter and its moons, Huygens was able to estimate the first value for the speed of light based on Rømer's numbers: an astoundingly fast 220,000 km/s. This value is 26 percent lower than the actual value, but the main point was that it was finite, rekindling Huygens' interest in dioptrics and in Pardies' wave theory of light.[20]

Huygens began again, but this time on a simpler problem than extraordinary refraction in Iceland spar. He chose to tackle the problem of the imperfect focusing of spherical lenses that spreads out the ideal focal spot into an extended flare of light. Something similar happens when a bare overhead light reflects from the sides of a circular water glass. A dazzling filament splays across the bottom of the glass with a sharp bright cusp at the center. These bright filaments of light are called caustics, meaning *burning*, as in burning with light. The name goes back to Archimedes of Syracuse and his apocryphal burning mirrors that were supposed to have torched the invading triremes of the Roman navy in 212 BC. By the time Huygens began to study the formation of caustic curves, he had already completed his theory of evolutes as they pertained to the pendulum clock. Evolutes were now the perfect tool with which to study caustics, because in geometric optics, light is viewed as a set of parallel rays that impinge upon an interface. The interface deflects the rays, after which each ray travels again in a straight path, but at differing angles. The envelope of all the deflected straight lines, the evolute, is where the rays pile up to create a bright streak—the caustic. This is a beautiful example where the mathematical evolute makes itself visible in a very dramatic way in everyday life. The caustic of a circular mirror is shown in Figure. 1.5.

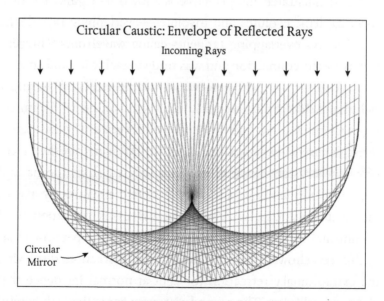

Figure 1.5 Caustic formed by parallel rays reflecting from a circle.

Huygens made rapid progress over only a few weeks. First, he connected the visible caustic evolute to its invisible involute. But what was this involute in the case of a spherical lens? It was a surface that was perpendicular to every ray—it was Pardies' wavefront! In addition, there is not only one involute, but also a continuous succession of involute curves that can be viewed as transforming one into the next as the wavefront propagates. Next—and this is the moment of genius that challenges our ability to understand how such leaps are made—Huygens realized that each involute was itself the tangent curve of innumerable little spherical wavelets originating from the preceding involute curve. Here is Huygens' principle: each point on a wavefront emits a wavelet that travels in all directions, yet only along the tangent curve do all the impulses add up and reinforce each other to create the new wavefront, shown in Figure. 1.6. Huygens' own drawing that first illustrated his principle is shown in Figure. 1.7 for a spherical lens. Evolutes and involutes—overlapping rays define caustics and overlapping tangents define wavefronts. Huygens had made the connection and was ready to tackle Iceland spar.

On August 6, 1677 Huygens wrote in large letters on the upper right-hand side of his notebook "EYPHKA, I found it!"[21] mimicking Archimedes. He had made two discoveries. The first came through careful experimental work in which he polished faces of calcite crystals along planes that differed from the natural growth planes, and he was able to isolate the symmetry axis of the calcite crystal. Light traveling parallel to this axis experienced no unusual refraction, while that light traveling perpendicular to this axis showed the maximum double refraction, displaying the extraordinary refraction for light at normal incidence that violated Snell's law. The second discovery came through careful

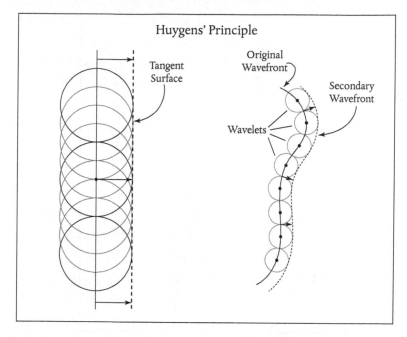

Figure 1.6 Huygens' principle. Every point on a wavefront is the source of a secondary spherical wavelet. The tangent curve to all the wavelets is the new wavefront.

deduction as Huygens determined that light traveling as waves in the crystal traveled with different speeds in different directions relative to the crystal symmetry axis. In other words, the little wavelets of his new principle, when emitted from a point in the crystal, traveled outward as an ellipsoid instead of a sphere. And when he constructed the tangent curve to the ellipsoids, he saw that it produced a plane wave that moves through the crystal at an angle—the extraordinary refraction that violated Snell's law, shown in Figure. 1.8. It also produced a ray that was no longer perpendicular to the wavefront, hence violating Pardies' third principle. In part, this is the genius of Huygens. Pardies' principle

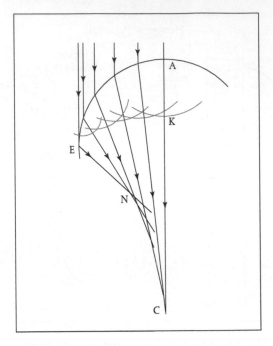

Figure 1.7 Huygens' caustic for a spherical lens. Rays incident on the lens far from the optic axis are refracted more strongly than rays close to the optic axis. This causes spherical aberration that smears out the focal spot. The envelope to all rays forms a caustic curve.

of normal propagation helped Huygens find his wavelet principle. But then his wavelet principle took precedence, and he concluded that the extraordinary rays in Iceland spar were not perpendicular to the wavefronts. A first principle helps establish a deeper and more general second principle that then overturns the first, showing it to have been a special case.

The story of Huygens' discovery of his principle is an important part of the bigger story of optical interferometry. It tells how a keen analytical mind conceived of the superposition of rays or wavelets that generated optical phenomena that can be

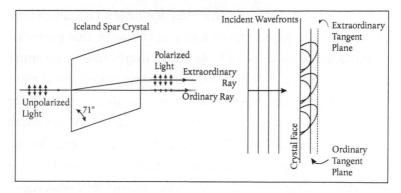

Figure 1.8 Iceland spar displays double refraction. One ray refracts in the ordinary way, but the other does not. The refractive index is a function of the crystal orientation and also of the polarization of the light.

experienced and investigated in the laboratory. However, the action of all the little wavelets to create the tangent curve was not yet the theory of interference. In Huygens' view, his wavelets were not the harmonic waves we think of today, but instead were impulses like shock waves. Huygens' wavefronts were singular surfaces, not continuous-valued functions that can be added in phase or out—that would have to wait for Thomas Young a hundred years later.

Young's Waves

Among the duties of a lecturer of the Royal Institution were experimental demonstrations illustrating new concepts. Davy and Faraday were famous for their public performances, played with flare to great dramatic effect, often involving small explosions or blinding flashes of light. For Young to illustrate his principle of constructive and destructive interference, he built a shallow tank of water with a glass bottom and a light source that projected up through the tank onto a large screen overhead that the entire audience could observe.[22] In the tank, Young placed metal wires that vibrated up and down, producing expanding circular water ripples. When the projected light was refracted by the water ripples, it produced wavelike shadows on the screen. Young had invented what is today called a "ripple tank," used almost universally across the world in physics lecture demonstrations for introductory physics students. With this device, he could form two circular wave patterns that overlapped, creating regions of constructive and destructive interference easily visualized by everyone. He later reproduced illustrations of these effects, shown in Figure. 1.9, in his *Course on Natural Philosophy*, published in 1807.

Young's preparations helped influence his own developing ideas of light and interference as he went beyond the physics

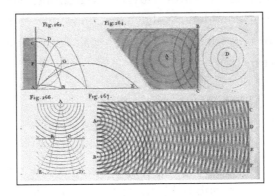

Figure 1.9 Reproduction of a part of Young's Plate XX showing reflection, diffraction, and interference as seen in a ripple tank.

T. Young, *Course on Natural Philosophy* (1807), PLATE XX, p. 777, https://openlibrary.org/books/OL7064835M/

of light to encompass all that was then understood in natural philosophy and the mechanical arts. Though not prone to exaggeration, Young mentioned to a friend that the nine months he spent preparing for the lectures, and then delivering them over the first half of 1802, was an effort that would likely kill him if ever repeated.[23] Despite his efforts, the fruits of this labor were mixed. Whereas Humphrey Davy and Michael Faraday were exciting and popular lecturers, giving public performances that were acclaimed as fine successes, Young's lectures were slow and plodding, difficult to understand, and lacking dramatic flair. Where Davy's lecture halls were filled, Young's were nearly empty. Nonetheless, Young was accomplishing something significant, establishing a foundation for natural philosophy that would last through much of the new century. Though they did not know it, the attendees at Young's lectures were seeing new discoveries almost daily.

Despite their lackluster delivery and attendance, Young's lectures continued in the first half of 1803, but by then Rumford had departed, and the Institution leadership had grown impatient with Young's poor performance, so he was relieved of his position. It is likely that Young himself was relieved to be out from under the responsibility of lecturing. Yet the transformation wrought on Young by this experience was impressive. By reading so widely, combined with the practical skills he had gained by preparing lecture demonstrations, he had become a first-rate natural philosopher of remarkable talent and skill. He had never lost sight of his key interest in the physics of light, and now he had the mathematical tools and the experimental ability to pursue his ideas in earnest. That year he won the Bakerian lectureship for the third time, to be delivered at the Royal Society in November of 1803, for which he chose the topic of physical optics. Free from his Institution lectures, he prepared himself for this momentous lecture.

Young concentrated on the phenomenon of diffraction. His goals were twofold: first, to use his new ideas of wave interference to explain observed diffraction effects, and second, to use diffraction as experimental proof of the wave nature of light.[24] Diffraction had first been observed and named by the Italian astronomer and mathematician Francesco Maria Grimaldi (1618–1663) when he measured the changing size of shadows as a function of distance behind illuminated opaque obstacles that were accompanied by rippled variations of brighter and dimmer intensity bands. A century later, in 1786, the American astronomer and inventor (and first director of the US Mint) David Rittenhouse (1732–1796) made the first diffraction instrument (a diffraction grating) when he strung an array of 50 hairs between the fine threads of two screws and passed a beam of collimated light through the structure. The array of hairs formed a diffraction grating with a density of 100 lines per inch that deflected the beam by 2 mm to each side at a distance of 1 m away from the grating. Others began to study diffraction effects and observed what were known as "internal fringes" that are dim bands of light appearing inside the shadow behind a solid thin object.

As he prepared for his upcoming Bakerian Lecture, Young covered a window with an opaque screen in which he pierced a small hole. A mirror located outside the window captured the sun's rays and projected them through the hole to create a thin beam of light that transmitted across the dark room to the far wall. Young then fashioned a very thin opaque screen, only one-thirtieth of an inch wide (a little less than a millimeter) that he positioned in the narrow beam of light. By observing the light patterns on the far wall, or on white cards that he could move closer to the opaque object, he was able to study the effects of light scattered from the screen edges. Of particular interest were the interior bands of bright and dim light that appeared on the white card where the shadow should have been most dark. The critical step in his investigation came when he used another card to block the light passing to one side or the other around the thin screen, and the intensity bands disappeared, reappearing when he removed the side obstruction. Only when light from both sides of the obstruction overlapped were the fringes observable—when light took two separate paths and interfered! To make sure that the fringes truly disappeared by blocking half of the light, and were not simply made too dim for him to see, he repeated the experiment by reducing the overall light intensity by a factor of 10 or 20 and continued to observe the interference effect.[25] For Young, this was conclusive evidence for his wave theory of light and for the principle of constructive and destructive interference. The description of this experiment was the centerpiece of his Bakerian Lecture delivered in November 1803 to the Royal Society and published the following year in the Royal Society's *Proceedings*. In the paper, he no longer referred to the *principle* of interference but rather to the *law* of interference.

Over the next four years, Young worked to consolidate the lectures he delivered at the Institution along with his studies of light interference, publishing his major work on natural philosophy as a two-volume set *Course of Lectures on Natural Philosophy* in 1807. It was the most comprehensive exposition of the state of natural philosophy in its day, a task for which Young was particularly well suited because of his polymathic genius. It contained his study of elasticity in which he defined the compressive modulus of materials, known today as "Young's modulus." It contained an estimate for the molecular size of water, based on an analysis of surface tension, nearly a hundred years before the atomic nature of matter was accepted. It contained a description of heat as a type of motion rather than as the substance caloric. It explained kinetic energy as the product of mass times velocity squared, and he called it "energy" for the first time rather than using Leibniz' outdated term *vis viva*. It contained a unification of light and heat by comparing thermal radiation to light radiation through space. In short, Young's lectures provided the cornerstone for the development of physics through the first half of the nineteenth century and was only eclipsed more than 50 years later by the famous *Treatise on Natural Philosophy* published of William Thomson (Lord Kelvin) and Peter Guthrie Tait in 1867.

The first volume of Young's lectures contained a refined version of his interference experiment in which he replaced his single narrow opaque obstruction with two narrow slits—now known as the famous "Young's double-slit experiment." One striking observation was the location of the brightest fringe at the very center of the diffraction pattern precisely where the shadow should be darkest. Young explained this very easily through the symmetric paths of the light rays causing constructive interference at the center of the diffraction pattern. This simple observation of the bright fringe at the center of the shadow is similar to the "spot of Arago" that would become the crucial point of the eventual acceptance of the wave nature of light 20 years later (see Chapter 2).

Young also became aware of the conditions for coherence, recognizing the importance of an illuminated pinhole as a single coherent source of light that subsequently could be split and recombined along slightly tilted paths. As the waves crossed at shallow angles, maxima and minima occurred in the light intensity. This was an important step beyond standing waves and the colors of films, moving into the problems of overlapping waves in space. His earlier vague thoughts on the superposition of waves now became focused on the path lengths taken by rays of light. Most importantly, he identified the conditions when maxima and minima occurred in the wave interference when two rays crossed at small oblique angles. This case was mathematically different from, and more mathematically complicated than, the standing waves in thin films. But he recognized that if the relative path length difference of the two rays were an integer number of wavelengths, then an intensity maximum occurred, while if the relative path length difference were an odd half-integer, then an intensity minimum occurred, as illustrated in Figure. 1.10 from our modern perspective.

On his part, Young should have been thrilled with his success, being at the pinnacle of his career as a physicist. However, he had challenged the Newtonian theory of the particle nature of light in England where Newton was almost as authoritative as God, and there were consequences. From the time of his first paper on the wave theory of light, through the publication of his third Bakerian Lecture, Young was subjected to increasingly personal attacks on his theory as well as on his character. The attacks were anonymous, but launched from the newly influential *Edinburgh Review*, casting doubt over Young's theory and experiments.[26] The reviews had little substance, other than diatribe, but they paint a picture of the general resistance to new ideas.

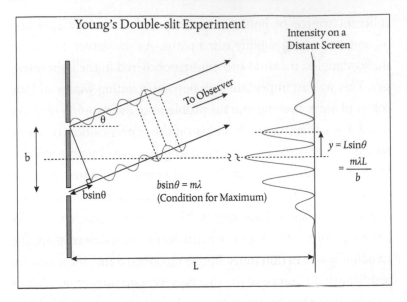

Figure 1.10 Young's double-slit experiment consisted of two narrow slits separated by a distance b illuminated by a distant point source of light and detected on a distant screen placed on the far side of the slits. When the path difference bsinθ between the rays emanating from the two slits at a given observation angle θ is equal to an integer number of wavelengths mλ, the peak and troughs line up and interfere constructively to produce an intensity maximum on the screen at that angle.

Despite their lack of merit, the reviews did touch on Young's weakness for being too qualitative. He was foremost a thinker—literally a natural philosopher—rather than a mathematician or experimentalist. In contrast, Newton and Huygens were each mathematicians and experimentalists in equal measure. In addition, Young's law of interference was based on the deflection of light by the edges of an opaque screen, using a ray theory of light waves in which interference effects occurred between rays rather than from the broad expanse of light emitted from the illuminated apertures. Despite embracing the wave nature of light, Young did not embrace the fundamental aspects of Huygens' principle. He believed that Huygens' secondary wavelets would cause light to disperse rapidly, preventing the formation of the thin beams of light that he used in his own experiments. Therefore, Young was a transitional figure in the development of the wave nature of light. He stated the law of interference for the first time in its clearest form, but his ideas and demonstrations were viewed more as intriguing than conclusive. The concluding steps toward acceptance of the wave nature of light would be taken a few years later in France by a builder of roads and bridges—someone who would build a theoretical edifice that stands to this day.

2

The Fresnel Connection

Particles versus Waves

> ... neither the corpuscular nor the undulatory ... will furnish that complete and satisfactory explanation of all the phenomena of light which is desirable.[1]
>
> John Herschel, *Treatise on Light* (1830)

Particles and waves are the yin and yang of physics. Throughout its history, physics has grappled to reconcile these two extreme views of dynamical systems. On the one hand, individual particles execute trajectories impelled by invisible forces, colliding sometimes with other particles at discrete points in space and time. On the other hand, continuous media like ocean waters move collectively in great undulations that spread out across vast distances. Particles are intrinsically local—waves are intrinsically non-local. Yet from the earliest days of physical thought, beginning with Democritus and continuing through Aristotle to the Enlightenment, it was suspected by some that the two were somehow linked—that waves emerge from media composed of unseen particles whose mutual small actions and reactions against each other combine to form the greater motions.

The conflict between particles and waves has always been most acute in the physics of light. A succession of wave-particle debates

Interference. David D. Nolte, Oxford University Press. © David D. Nolte (2023).
DOI: 10.1093/oso/9780192869760.003.0002

concerning the nature of light has engaged some of the greatest scientific minds in the history of physics. The "first" debate in the late 1600s pitted the wave ideas of Pardies and Huygens against Newton's corpuscles, and Newton was the victor, forming a seemingly impenetrable edifice surrounding the corpuscular character of light. But Euler, who codified most of what we today call Newtonian physics, did not agree fully with Newton's corpuscles of light and adopted an intermediate position that created cracks in Newton's edifice—cracks that Young later exploited to propose the wave nature of light as the cause of diffraction and interference. This launched the "second" debate in the early 1800s, as those who supported the emission of light particles—known as the emissionists—waged a minor war against those who supported the wave properties of light—known as the undulationists. The emissionists won many of the battles, such as the discovery of the polarization of light that initially struck what seemed a fatal blow at the heart of undulation, but they ultimately lost the war.

The Sun Setting on Luxembourg Palace

By the time Pierre-Simon Laplace had finished the fourth volume of his masterwork *Méchanique céleste* in 1805, he had completed a Newtonian synthesis of all known interactions, treating both light and massive particles through central forces. From that moment, Laplacian optics became a pillar of French science. It was also one of the first unification theories in the history of mathematical physics—though it would not stand the test of time. The elegance of the theory was the best argument in its favor, and Laplace—ever convinced of the superiority of his own views—was committed to promoting his theories and establishing his legacy. To further his ends, Laplace enlisted his colleague Etienne Malus of the Egyptian campaign and of the society that Berthollet had established in Arcueil, to provide supporting evidence for his theory of light.

Through careful experiments on the reflection of light from opaque materials, and with a keen eye on experimental uncertainties, Malus was able to demonstrate the superior predictions of Laplace over an alternative theory by the British chemist and physicist William Hyde Wollaston (1766–1828). As he continued his optical researches beyond the whims of his mentor, Malus returned to ideas he had first conceived during long treks across Egyptian deserts and while surviving dreary hours of confinement with the plague in the Levant. He proved a proposition on the mathematical properties of families of rays reflecting or refracting from non-planar surfaces, now known as Malus' theorem. Several decades later, William Rowan Hamilton would use Malus' theorem to create a general theory of ray optics that was so fundamental that he later applied it to trajectories of particles, creating the field of Hamiltonian physics.

The prize committee of the *Académie des Sciences*, encouraged by Malus' success applying Laplacian optics to diverse problems, announced a prize for the physical and mathematical explanation of double refraction—aiming an arrow at the heart of Huygens' wave theory of light developed 150 years earlier to explain double refraction in Iceland spar. If the emissionists could use Laplace's theory of light to explain the effect, it would help put a stop to the resurrection by Young of the annoying wave idea. Malus was ideally positioned to compete for the prize, and he began to play with crystals of Iceland spar as he developed an emissionist approach to double refraction.

In 1808 Malus occupied a house on the Rue d'Enfer on the southeast side of Luxembourg Park in Paris where his study looked out across the manicured gardens to the stately Luxembourg Palace. Late one fall day, the setting sun's rays were reflecting from the palace's windows in a brilliant blaze of orange and gold. He looked through a crystal, expecting to see the usual double image of the Palace, which he did. But what he did not expect to see was that the brilliant sunlight reflecting from the Palace windows could be seen in only one of the two images. The windows of the second image not only were not bright, they appeared as extraordinarily dark panes! In amazement, he studied the effect as the sun set and night fell, turning the crystal this way and that to try to understand the phenomenon.

As the Palace lighted its rooms, the usual double images of illuminated windows reappeared. Malus lit candles in his own study and in a flash of insight decided to look through his crystal at the reflection of candlelight off the still surface of water in a basin. When the reflection was viewed at about 36 degrees, he observed the same effect as the sunlight off the windows of Luxembourg Palace, as only one image of the candle appeared through his crystal. Being a natural experimentalist, he tried additional observational configurations. For instance, he placed the crystal between the candle and the water, and as he looked at the reflection, again nearing 36 degrees, one image was reflected while the other disappeared. Malus must have known that he had discovered an essential new property of light that no one had known before. Without waiting to put this discovery in his prize memoir, he delivered a paper before the *Académie* on December 12, 1808, announcing his discovery of light polarization by reflection.[2] Within the context of Newton's particles of light and Laplace's optics, Malus incorporated a "sidedness" to light particles and developed an elegant theory of double refraction based on oriented light particles for which he won the *Académie* prize in 1810. Malus' discovery was profound, but his theory was enmeshed in the wrong model for light. The phenomenon, as it is understood today in terms of polarized light waves, is illustrated in Fig. 2.1.

Word of Malus' discovery of polarized light spread rapidly, and already in 1809 Thomas Young realized that there was nothing in his wave theory of light that could explain the effect, because at that time he believed that light was like sound in air with only

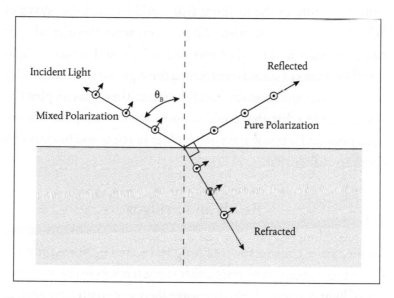

Figure 2.1 Polarization by reflection at the Brewster angle θ_B. The mixed polarization of the incident ray is converted to a pure polarization in the reflected ray. At the Brewster angle, the reflected and refracted rays share a 90-degree angle.

longitudinal modes. In a review of a memoir by Laplace describing Malus' work, Young wrote, "The same volume contains also an account of some highly interesting and important experiments of Mr. Malus, on the apparent polarity of light, as exhibited by oblique reflection, which present greater difficulties to the advocates of the undulatory theory than any other facts with which we are acquainted."[3] Later, in 1811, Young wrote to Malus directly to tell him "Your experiments demonstrate the *insufficiency* of a theory (that of interferences), which I had adopted, but they do not prove its *falsity*."[4]

The future of the undulatory theory of light was cast into serious doubt if even Thomas Young, its greatest proponent, faltered. The problem, the undulationists quickly grasped, was with the nature of the medium that could support the waves of light, the luminiferous ether. All previous wave theories of light had assumed that the ether was like a fluid, and waves of light were like waves of sound transmitted through the air in successive compressions and rarefactions of the ether. There was no physical science for an ether that could support waves with polarization, but a poor and unknown French builder of roads and bridges was about to change that.

Roads and Bridges

The French provincial town of Broglie, located in Normandy, is a strange focal point of physics and history. It is where the wave theory of light converged with the wave theory of matter, coinciding in space though separated by nearly 100 years in time. In 1924, the future seventh duc de Broglie, as a post-graduate student at the University of Paris, deposited a doctoral thesis that synthesized

Einstein's theory of the photon with Bohr's theory of the electron. Though the dissertation initially was greeted with little fanfare, Einstein mentioned de Broglie's work as a footnote in a paper read later that year by Erwin Schrödinger, who tracked down the thesis, taking both it and a mistress with him on a Christmas vacation to the Austrian mountains, returning a week later with a quantum wave equation for electrons. One hundred years earlier, while working and living at the de Broglie estate at the time of the second duc de Broglie, the architect Jacques Fresnel met the daughter of the estate manager. They married and had two sons, Louis and Augustin. But the French Revolution erupted, the duc decamped to safer territory, and the Fresnels moved to Cherbourg and finally, during the Reign of Terror, to the safety of the family village of Mathieu outside Caen where two more sons, Léonor and Fulgence, were born.

The second son, Augustin-Jean Fresnel (1788–1827), was neither the brightest of the four Fresnel siblings nor the most robust. By the age of eight he could barely read, lagging sadly behind even his younger brothers in school. Nonetheless, he had acquired a strange nickname among the village children, who called him "genius," because he had a knack for invention. For instance, as the children played war in the streets and fields, they fashioned toy cannons out of alder trees and staged mock battles. Augustin improved on the cannons, finding just the right trees to bore, and just the right projectiles to launch with just the right force, that the toys became real weapons. The children watched in awe and amazement at the stately arc of dirt-clod cannon balls before they smashed into the ground, spraying the combatants with debris. The village adults had to step in and ban the battles before anyone was seriously hurt.[5]

When Augustin's gifted older brother Louis was sent to the technical high school in Caen, Augustin tagged along to keep him company, and he slipped through the school nearly unnoticed. Yet the technical education resonated with Augustin, who was adept at mathematics and technical drawing, if not language and the humanities. When it was time for Louis to attend the new Ecole Polytechnique in Paris, the top technical school in France, Augustin tagged along again, this time placing a surprising seventeenth in the entrance exam out of an entering class of 300. He continued to do well, and after graduating he settled on a career in civil engineering, enrolling in the school for bridges and highways, run by the new Department of Bridges and Highways established by Napoleon after crowning himself Emperor in 1804. Civil engineering was a good career track in the new administration, with its emphasis on building a modern France. Unfortunately, building a modern France meant building bridges and roads in the most remote and backward parts of the country, like the notorious Vendee.

The Vendee was known for rebellion—rebelling against taxes and against conscription into the army while raising up homegrown militia who opposed the government. It took a small military invasion to finally subdue the province. The best defense by the government in Paris against another uprising was to integrate the area with the rest of France, connecting it by roads and bridges (which could also speed future troops to the area). The hapless engineer assigned to oversee this task was Augustin Fresnel. It was a terrible mismatch of talent and skills. Fresnel was socially awkward and disliked interacting with people, especially the uneducated and untrained laborers with their petty problems and squabbles. He also had a weak constitution and already had the early signs of consumption, which sapped his strength and his resolve to deal with day-to-day minutia. The work was tedious and unfulfilling, so he escaped, when he could, into books of mathematics and physics. Fresnel was captivated by technical questions and thrilled with the many topics of physics, one in particular catching his attention—the nature of light.

Fresnel collected all the books on physics and light that he could, reading and working out problems when he could spare the time. Unfortunately, his weakness in language prevented him from reading in any language other than French. Therefore, he worked in a near vacuum, teaching himself the details of the topic unaware of the broader efforts or contexts of international science, and certainly unaware of Thomas Young's work. Most French physicists at the time accepted Newton's particulate nature of light, but Fresnel had early insights that made him uneasy with the status quo. Perhaps it was his naiveté and ignorance of the arguments that allowed him to think more clearly than others, unclouded by given wisdom. His lonely isolation in the remote Vendee, and later in the south of France where he helped supervise the construction of a road linking Spain with northern Italy, helped in this respect, allowing him to see with fresh eyes the periodicity in the bright and dark fringes cast by diffraction into the shadows behind opaque objects.

Periodicity strongly suggests waves, and perhaps inspired by ripples launched by stones dropped into the calm canals beneath the bridges he was building, Fresnel conceived of interfering waves as the origin of the diffraction patterns. He considered how differing path lengths would allow peaks and troughs to build upon each other either in construction or destruction of amplitude. If light were made of waves, surely this could explain the bright and dark fringes of diffraction. He wrote up a simple letter, one of his first attempts at generating professional scientific work on his own, which contained broad speculations and general arguments rather than a concrete model, but it was a start. The next step was to find if his ideas had any merit.

Fresnel's uncle, Mérimée Fresnel, was a well-known artist and art restorer living in Paris. He moved in educated circles and knew many of the men of letters and science of the day, tending to know the up-and-comers more than the stodgy establishment. Augustin sent his tentative letter to his uncle, and Mérimée passed it on to André-Marie Ampere, an up-and-comer not yet well known, but certainly someone to watch. Ampere promised to take a look at the letter, but reading naïve ideas written by dilettantes in distant provinces is always awkward, and Ampere politely "lost" Fresnel's letter.[6] To Mérimée's credit, he had some feeling for the potential of his nephew, and he looked for other avenues to help him. One evening, he found himself seated at a formal dinner next to Francois Arago (1786–1853), a physicist and unlikely celebrity in French society, famous for his remarkable swashbuckling backstory.

Arago was only 20 years old when the *Bureau des Longitudes* selected him and his friend Jean-Baptiste Biot, freshly graduated from the Ecole Polytechnique, to extend Le Meridien, the line of longitude passing through Paris (the French alternative to the meridian passing through Greenwich England), as far south as they could manage. The *Académie des Sciences* in 1791 had defined the meter as one ten millionth of the distance along the meridian from the North Pole to the Equator. Therefore, the meridian was the fundamental basis of the new metric system of measurement.

The two friends set out in 1806, climbing mountaintops to tri-
angulate to the next location along the meridian and repeating
the process day after day and month after month for two years.
Finally, in 1808, as they reached the southern shore of Spain,
Napoleon launched an ill-conceived invasion of Spain, instantly
making Arago and Biot enemy agents in a hostile land. This same
invasion sadly took the life of Fresnel's beloved brother Louis who
was serving in the French artillery. Biot had the good sense to flee
back to France, but Arago had one last mountain to climb, on the
island Ibiza in the Balearic Sea off the coast from Valencia (see the
map in Fig. 2.2). The local authorities, growing suspicious of his
strange activities, arrested Arago as a French spy and threw him
in jail.

Fortunately, Arago was half Spanish, coming from the south-
ernmost region of France adjacent to Aragon (which was also the
origin of his family name), and spoke Catalan, the local dialect.[7]
His knowledge of the language and his familiarity with Catalonia
won over one of his jailors who helped him escape. He climbed
into a small fishing vessel and sailed following the prevailing
winds 200 miles southward to Algiers on the coast of Africa. He
introduced himself to the Dey of Algiers, who agreed to let him
embark on a boat the Dey was sending to Marseilles with two
lions and a troop of monkeys as gifts to Napoleon. The vessel
was already within sight of Marseilles when a Spanish corsair
intercepted them and took them as prize to Roses, Spain. Once
more imprisoned, Arago was able to have a letter smuggled out
to the Dey, informing him of the fate of his ship and gifts. The
Dey was furious and through diplomatic channels induced the
Spanish to free the ship and crew. Heading back across the Bay

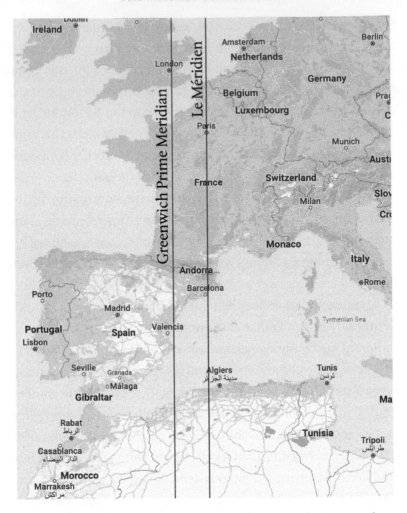

Figure 2.2 The meridians through Paris and Greenwich. Arago and Biot followed the Paris meridian down to the Balearic Islands of the Mediterranean Sea.

of Marseilles the ship was struck by a strong *mistral* that drove it relentlessly across the Mediterranean to the coast of Algeria, landing near Béjaïa 170 km to the east of Algiers. The captain of the ship received new orders, stranding Arago, a Christian in a Muslim land, with no means of transport. So, he traveled by land, disguised as a Muslim trader, across the desert and the coastal inlets, back to Algiers. Three years after Arago left France, the *Bureau des Longitudes* declared him dead and gave official notice to his family. Before they had time even to grieve, Arago reappeared as if by magic at the port of Marseilles on the second day of July 1809. Through all his troubles and travels he had held onto his records and equipment, which he now deposited at the *Bureau des Longitudes*, completing the Paris Meridien as far south as the Balearic Islands.

Arago's wild adventures and miraculous return from the dead made him a favorite of the salons of Paris. He was a gifted storyteller and enthralled his audiences, ensuring his place as a star of Paris society. His accomplishment for the *Bureau des Longitudes* also ensured his place in the society of scientists, and he was nominated to the *Académie des Sciences*. His election seemed assured because he expected to have the backing of Simon-Pierre Laplace, the leading physicist of France. Laplace had been a staunch supporter of Arago and Biot for their mission to extend the Meridien, and once back in France, Laplace continued to enlist Arago on more theoretical projects, such as his new program to provide the same rigor to Newton's theory of light that he had already provided for Newton's theory of gravity. But gradually, Arago's relationship to Laplace soured as he got caught up in political squabbling over the membership to the *Académie*.

Laplace was backing Siméon Denis Poisson (1781–1840), at that time 28 years old and already a considerable force in French science. Poisson was incredibly prolific during his career, publishing hundreds of papers across all the frontiers of mathematics and physics of the day. In 1809 most of this was in his future, but Laplace already saw in Poisson his intellectual heir, and they became very close. Arago was not ill disposed toward Poisson, with whom he had long shared mutual respect and friendship back to their days as students at the Polytechnique, but Arago developed a visceral dislike of Laplace, not for backing Poisson, but for being so arrogant and dismissive of anyone he viewed as inferior, which included almost everyone. Laplace did not hesitate to undermine Arago at any chance, interfering with the election where he could, which was a breach in the civil tradition of the *Académie*. Laplace's tactics backfired. He may have been held in considerable esteem by his colleagues, but he did not have their affection, and Arago was voted in with a landslide election.

Arago was pleased with the outcome, but also was bitter against Laplace, and even his friendship with Poisson cooled. He began to recognize the stranglehold that Laplace and the old guard had on French physics. Arago was young and idealistic, a product of the French revolutionary spirit against tyranny of any form, and he looked for ways to fight it. An odd outlet for this rebellious spirit came in the form of Newton's particulate theory of light. Laplace, with the help of Poisson, was in the process of codifying the principle, with the goal of making it unassailable, so Arago decided to assail it. In 1811 he published a short paper that highlighted several tricky problems with the emission theory, one of which was the problem of diffraction. The paper offered no resolution of these problems, but it heightened Arago's awareness of the obvious deficiencies of the notion of particles of light.

Therefore, it was a fortuitous moment a few years later when Mérimée sat down at the table next to Arago and casually mentioned his nephew's views on the mechanism of interference and its possible role in the formation of diffraction fringes. Arago immediately recognized the merit of the argument and inquired who this nephew might be, finding that it was a younger classmate of his from the Polytechnique, though they had moved in very different circles. Arago encouraged Fresnel's uncle, who wrote to Fresnel of the successful reception of his ideas. This must have given some comfort, providing a ray of hope to distract Fresnel from his road building.

French politics at that time was in constant turmoil. Napoleon had been deposed and exiled in 1814, replaced by an unstable royalist government, leaving the door open to his return. When he landed in the south of France in February 1815, an army of passionate supporters hastened to his rallying call as he began his march on Paris. Just as passionate were many royalists, including the usually meek and mild Fresnel, who left his post to join a hastily formed militia to try to stop Napoleon's drive north. The untrained royalists were no match for Napoleon's tactical genius, and their attempt to stop him failed. Fresnel returned to his post, downcast and vilified by the local citizens, who had embraced Napoleon's return, and he was placed under house arrest as an enemy of the state. Despite the turmoil, life during Napoleon's last hundred days remained civil, and Fresnel's supervisor arranged for him to return to his hometown of Mathieu where he could spend his suspension in familiar surroundings. Fresnel traveled to Mathieu through Paris, where he stayed with his uncle Mérimée and met Arago for the first time.

Arago and Fresnel were opposite poles. Arago was vibrant, forceful, an ardent radical, self-possessed, robust, charismatic, and confident. Fresnel was timid, a staunch royalist, unsure of himself, sickly, and searching for meaning in his life. Yet their meeting sparked a lifelong friendship. Arago saw in Fresnel an inventive and creative soul with admirable technical skill. Fresnel saw in Arago someone with a sincere interest in his ideas and a quick mind who could immediately grasp his meaning. They met several times, lost in deep discussions that must have thrilled the intellectually starved Fresnel. Arago was also in a position to bring Fresnel up to speed with the current state of knowledge, since Fresnel's work in the provinces had cut him off from the mainstream. It was at this time that Arago first informed him about Young's experiments from 10 years earlier and gave him a copy of Young's 1804 Royal Society paper (Young's Bakerian Lecture of 1803) to take with him to study.[8] Fresnel, always weak at languages, was not able to read the paper in any detail, but by the time he arrived home in Mathieu, Fresnel had become convinced of the importance of the problem of diffraction and the validity of the wave theory, and he was determined to prove it.

For people of the mind, house arrest is not necessarily a bad thing. Galileo's conviction of heresy at the hands of the Roman Inquisition and subsequent house arrest provided him uninterrupted time to finish his lifelong work on motion, enabling him to publish his *Two New Sciences* in 1632. Without the discipline of house arrest, it is possible that the world would never have seen the completion of his work. For Fresnel, his return home was an escape from the distractions of road building, freeing him to pursue his interest in the physics of light.

The local locksmith in Mathieu was a jack of all trades who helped Fresnel set up a dark room furnished with optical mounts and experimental tools. The light source for his optical experiments was the sun shining through a small hole drilled in a metal plate and filled with a bead of his mother's honey to focus the light. Fresnel strung a thin wire in the light beam to act as an opaque element to diffract the light, and he placed a white sheet against the far wall where he could see the dim diffraction fringes. At this stage, his approach and his measurements were not much better than Young's, merely confirming the general properties of the diffraction pattern. The dimness of the spread-out fringes, their fuzziness, and their narrowness prevented accurate recordings using a ruler. Frustrated and stumped, Fresnel likely stood in front of the white sheet pondering how to improve the accuracy, and when he looked directly back at the wire in the sun beam he would have seen the clear and sharp diffraction fringes focused on the retina of his own eyes. The fringe brightness and contrast far exceeded the barely perceptible fringes cast on the sheet.[9] Clearly, the human eye was a far superior imaging system and photodetector than diffusely scattered light from a white screen, if he could only put numbers to what he saw in his head.

Fresnel had an epiphany. With painstaking care, he fashioned a triangular mesh made of fine threads spaced with the greatest accuracy that he could achieve, positioning this mesh in the midst of the diffraction pattern. He then looked with his eye back toward the beam of light, focusing on the mesh. The fringes and mesh both remained in focus, allowing him to compare the positions of the fringes against the locations of the threads. It was then just simple geometry to convert the relative positions to accurate values for the fringe locations. The accuracy of this approach located the fringes to within 200 microns. For a metal wire thickness of half a millimeter, and a viewing distance of several meters, the accuracy in the location of each fringe was better than 10 percent. He was also able with his sensitive eye to measure out to very large fringe numbers, increasing the accuracy of the fringe spacing. This approach exceeded anything Young had accomplished, and it agreed with the wave theory predictions to better than 1 percent. He wrote up his findings in an article that he sent to Mérimée who passed in on to Arago for submission to the *Académie*.

Unfortunately for Fresnel the royalist, Napoleon was defeated at Waterloo on June 18, 1815, restoring the Bourbon monarchy, which Fresnel desired, but ending his house arrest, which he did not. He was recalled by the Bureau of Roads and Bridges to his post, even receiving a commendation for his brief stint in the militia that had opposed Napoleon's return. But the honor was drowned by his dread of returning to road building. He petitioned for an extended leave of absence to stay with his mother in Normandy as it was inundated by Prussian occupation forces. The Bureau was not eager to let its employees lie idle, and was about to reassign Fresnel, when Arago intervened. He had been selected to referee the paper submitted by Fresnel to the *Académie*, and he was thrilled at what he saw. The thoroughness, cleverness, and precision of Fresnel's observations were unlike anything he had seen, and it was one of the most forceful arguments in favor of the wave nature of light. Arago went through back political channels to have Fresnel reassigned to Paris, at least temporarily, so that he could work directly with him in the Observatory facilities, repeating the measurements and seeing where the results would further lead. France was again in chaos, with roads a low priority, so Arago's gambit succeeded. Fresnel returned to Paris in the spring of 1816.

Frankenstein Summer

The summer of 1816 across Europe was no summer, with end-less days of cold rain that prevented sunshine from casting the light Fresnel needed to demonstrate his findings to Arago. The entire Northern Hemisphere had been enveloped in the gloom of volcanic ash from a giant eruption the year before in Indonesia. Snow lay on the streets of New York in June, while storms raging over Lake Geneva kept a small group of British tourists indoors for days as they concocted ghost stories to pass the time.[10] One of the party was Mary Wollstonecraft, traveling with her lover and future husband Percy Shelley, accompanied by Lord Byron and John Polidori. Mary's ghost story was about a scientist who used the electrical arts, newly discovered by Alessandro Volta, to reanimate a cadaver. John's contribution detailed the macabre tales of a blood-drinking British nobleman. On their return to England, Mary wrote and published her epic *Frankenstein or The Modern Prometheus*, while John's short fiction *The Vampyre* became the first of the romantic vampire genre of fiction.

The failed summer helped Fresnel forge his own creation, with encouragement from Arago, as he expanded on his earlier letter to show the detailed agreement between the wave theory of light and his accurate measurements of the fringe spacings. Most importantly, Fresnel developed his theory of the "efficacious ray" that, for the first time, went beyond all previous theories of light scattering from obstructions. Previously, both undulationists and emissionists had viewed light scattering as a mechanical interaction that occurred at the exact edge of the opaque obstacle. For undulationists the edge created secondary outgoing circular

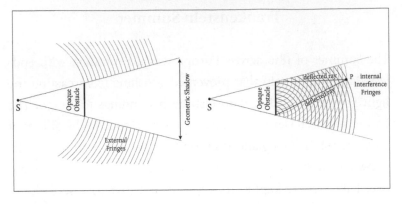

Figure 2.3 Scattered rays and waves from a source S on an obstruction that casts a classical shadow. In the old diffraction theory, the edges of the obstacle deflect rays, or generate circular waves, that extend into the shadow region.

waves, while for the ray theorists the edge deflected the rays, as in Fig. 2.3. But Fresnel, an undulationist and accurate experimentalist, recognized that the observed internal fringes that occurred in the "shadow" of the obstruction only crudely agreed with the theory. He got much better agreement if the source of the secondary waves was located a small distance laterally from the edge. At first, this seemed like a classic "fudge factor" with little scientific value, until Fresnel realized that the small distance was not arbitrary but corresponded to an optical path difference of half a wavelength. This was the crucial epiphany that launched the modern theory of diffraction. He recognized that the free space to the side of the obstruction produced regions of waves that alternated their optical path lengths by integers of half wavelengths, thereby exactly canceling out, except for the very first region, which was the source of his efficacious ray. Today, these regions are called Fresnel zones, shown in Fig. 2.4. Although Fresnel could not yet

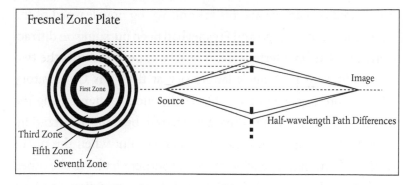

Figure 2.4 Fresnel zone plate. The path difference from source to image points across each zone is a half wavelength. Each odd-numbered zone contributes constructive interference at the image point.

devise a way to precisely add multiple waves of differing phases, he had taken the first step toward establishing a rigorous theory of diffraction.

The same failed summer that created Shelly's *Frankenstein* and Fresnel's efficacious ray also created a famine across the rural regions of France, and Fresnel was called to duty overseeing work houses in Brittany, putting the starving farmers to work on roads and bridges for meager wages. In small periods of time that he could take that were free from famine relief, Fresnel continued working on his new theory of zones and the addition of an arbitrary number of waves. This mathematical problem remained the crucial obstacle to a more accurate theory of wave interference and diffraction. It was also the crucial breakthrough that he needed to finally step out from behind Young's shadow.

In 1816 Thomas Young received a historic visit from Arago in London, and the two found much common ground for deep discussions. Young had been mostly unaware of Fresnel's work in the field of wave optics, since French optics at that time was dominated by Laplace and his school. Young now learned of the experiments that Arago and Fresnel had been pursuing on diffraction as well as on interference. A crucial experiment that the two had performed in Arago's laboratory at the Paris Observatory was the interference of two polarized pencils of light, using the methods of Malus. When the waves shared the same polarization, interference fringes were clearly observed. But when the polarization of one of the pencils was made perpendicular to the other, the fringes disappeared. Fresnel used these experiments to help develop his ideas, but had not yet hit on the essential meaning behind the non-interference of orthogonally polarized light.

During Arago's visit, none of the English scientists were impressed by what Arago told them, and he may even have mistakenly given them the impression that Fresnel was merely repeating Young's work. Furthermore, with the exception of Young, few agreed with the wave theory of light, and even the novel results that Fresnel had generated were used as further proof of particle and ray theories.[11] When Arago reported back to Fresnel about the weak British reception, Fresnel was crestfallen and briefly considered giving up all his researches as a "stupid plan ... to acquire a small bit of glory."[12] This was his dark frame of mind as he was called back to his duty in Brittany that he could not stand.

However, Arago had only just returned to France when Young realized that light must have a motion transverse to its direction of propagation! In this way, light could be decomposed into two transverse polarizations and hence explain the formation of just the two images in Iceland spar. He quickly sent off a letter to Arago in January 1817 telling him of this idea. When Arago showed Young's letter to Fresnel, he instantly understood what it meant and went a step further. While Young was still thinking of transverse *and* longitudinal waves, Fresnel combined the idea of transverse waves with the experimental results that he had obtained with Arago on the non-interference of orthogonal polarizations. With this, Fresnel now proved that light could be *only* transverse, with no longitudinal component at all.

Meanwhile, Arago brought his growing political weight to bear on the head of the Department of Roads and Bridges, pressing him to reassign Fresnel to Paris, still to oversee roadwork, but at least to bring him in from virtual exile in the provinces. His efforts succeeded, and Fresnel gratefully returned to Paris in the fall of 1817. With renewed faith and ample time, Fresnel concentrated on the remaining central problem of adding an arbitrary number of waves together to produce a resultant intensity. This problem used the principle of superposition which had been discovered a hundred years before by one of the bountiful Bernoullis of Basel.

Daniel Bernoulli and the Principle of Superposition

The 25 years from roughly 1730 to 1755 represented the golden age of "rational mechanics."[11] This was the post-Newtonian period when rapid progress was made on the mechanics of rods, beams, strings, membranes, and fluids. This quarter century also coincided with the development of the first differential equations of motion and statements of the first *partial* differential equations. Key advances on these topics were published by Daniel Bernoulli on compound motions[14] (1733) and hydraulics (*Hydrodynamica* 1734), by Euler on partial differentials (1744) and infinite series (1748), by D'Alembert and Euler on heavy strings (1747–1748), by

Daniel Bernoulli (1753) on the principle of superposition,[15] and Alexis Clairaut's (1713–1765) treatise on lunar tables 1754) which contained the first seeds of Fourier analysis. The speed of progress and the productivity of the participants during this period were astounding. It stands out in the history of physics as one of the rare periods of explosive growth of new concepts that developed rapidly into the forms we recognize today. The most famous equation of all, what we call Newton's second law, $F = Ma$, is actually due to Euler (1750) and dates to this golden age rather than to Newton himself (who never wrote the equation down).

Surprisingly, the most bitter technical controversy that emerged from this fruitful period arose from the simplest mechanical system—the vibrating string of a musical instrument. From the time when Brook Taylor had written down a sinusoidal equation for the fundamental mode of a vibrating string in 1713, many had searched for the "equations of motion" of the string whose solution would yield not only the fundamental mode but also the more complicated time-varying motions that were known (and heard) to occur. The aging Johann I Bernoulli came very close to defining the equation of motion in 1742 when he was 75 years old and only six years before his death. He considered a heavy string to be a series of point masses coupled by elastic but massless strings. He used advanced principles of pair-wise forces and boundary conditions, and he derived the equations for each point mass, but he did not quite make the connection to a differential equation.

The breakthrough to the first differential equation of motion was made by Jean le Rond D'Alembert (1717–1783) in 1743 for a heavy hanging cord by considering the case of N point masses. D'Alembert was an influential physicist and *philosophe* of the

French Enlightenment, co-editing Diderot's famous *Encyclopedia*. His early contributions to Newtonian physics stand out in particular through what is today called "D'Alembert's principle" which laid the foundation for Lagrange's later mechanics as well as Einstein's general theory of relativity.[16] Three years after demonstrating the case of N point masses, he tackled the vibrating string and wrote down the first partial differential equation for a continuous mechanical system—his famous wave equation. He found the general solution that is taught in every introductory physics course today, namely $\psi(x, t) = f(x - vt) + f(x + vt)$.

While D'Alembert was developing the first theory of partial differential equations, he was in a close correspondence with Leonhard Euler. At first, they were on good terms and exchanged cordial letters. As soon as D'Alembert published his solution to the motion of a taught plucked string, Euler responded with a paper of his own, disagreeing with D'Alembert on what the most general solution should be. The disagreement was highly specialized, relating to whether a solution could have a "kink" in it or not. Euler argued that at the moment the plucked string is released, the string would have a kink at the location where it was being plucked. Euler's point of view was mechanical in nature rather than mathematical, but it stemmed from good common sense. D'Alembert, on the other hand, was appalled and rejected Euler's approach, claiming that the derivatives in the wave equation would be violated at a kink, and hence any mechanical curve with such a kink was not allowable. For such a seemingly arcane point, the argument devolved into bitter diatribes that went back and forth for years. Euler, for his part, moved on, but D'Alembert held onto the feud literally for decades, self-publishing pamphlets that carried his polemics against Euler into the streets of Paris.

As Euler and D'Alembert argued in public, Daniel Bernoulli watched with dismay. Daniel was Johann I's second son, born in 1700 Groningen in the Netherlands when Johann was yearning to return to Basel, but was prevented, because the chair in mathematics in Basel was occupied by his older brother Jacob. Five years later, Jacob succumbed to consumption, and Johann gratefully took up his coveted post. Once back in Basel, Johann decided that one mathematician in the family was enough, and he arranged for his sons to take up more fruitful occupations, deciding on a business career for Daniel. However, mathematics must run in the genes, because both Daniel and his older brother Nicholas II gravitated to mathematics (as did half a dozen other Bernoullis), and the brothers were offered dual chairs of mathematics at St. Petersburg in 1725. Daniel thrived in St. Petersburg, especially after Leonhard Euler (his father's most able student), arrived in 1727. Daniel's brother Nicholas unfortunately died, but Daniel and Euler worked closely together for five years as an intellectually vibrant and symbiotic team. Daniel provided physical insights and experimental demonstrations while Euler provided the rigorous mathematics.

Despite his insistence on being a mathematician, Daniel was at heart a physicist. Around 1733, as Daniel was contemplating returning to Basel, he studied the normal modes of a compound double pendulum and discovered that all general motions were composed of combinations of fundamental modes. His subsequent studies of heavy cords led him to discover higher harmonics of Brook Taylor's fundamental sinusoid. In the limit of a string with a continuous mass line density, Bernoulli understood that there is an infinity of normal modes. Based on this, and on his idea of composite motions, he took the bold step to declare in a paper in 1753 that *all* motions of a vibrating string could be expressed as an infinite series of sinusoidal harmonics.[17] Daniel proposed the infinite series

$$u(x, t) = \sum_{n=1}^{\infty} a_n \sin\left(\frac{n\pi x}{a}\right) \cos\left(\frac{n\pi c}{a}(t - t_0)\right)$$

This was the definitive statement of the principle of superposition, and Daniel Bernoulli holds the priority of making this claim, extending back to his work on normal modes of the double pendulum in 1733.

One might assume that Euler would have approved of this principle, because he had already discovered, in 1744, that trigonometric series could be used to describe certain functions. This was a major step beyond the work of Taylor (1713) and Maclaurin (1742) who expressed functions in terms of polynomials. The problem with polynomials is that they diverge when the argument gets large, and any finite series can only approximate a function on a restricted domain. With trigonometric series, on the other hand, each sinusoidal function is bounded, and so a finite truncation does not diverge anywhere on the domain of the argument. This was a major improvement when one wanted to make only the smallest number of calculations to approximate a function. Unfortunately, Euler did not know how to find the coefficients except in a few special cases.

Daniel Bernoulli, as he expressed his principle of superposition, was in the same predicament as Euler, because he could not calculate the coefficients to prove it. For this reason, Euler rejected Bernoulli's assertions about the vibrating string just as D'Alembert had rejected Euler's. Daniel, caught in the middle, quipped that D'Alembert's and Euler's solutions constitute "beautiful mathematics but what has it to do with vibrating strings?"[18] The feud over the vibrating string expanded during the following years and picked up a new antagonist when a young Joseph-Louis de Lagrange (1736–1813) entered the fray in 1759 with an uncharacteristically flawed memoir on the problem.[19] Polemics flew back and forth among the combatants: D'Alembert, Euler, Daniel Bernoulli, and Lagrange. The arcane battle centered on the meaning of what a mathematical function can or cannot be. Polemics and personal feelings aside, the controversy holds a central position in the early history of mathematical analysis because it helped to expand the concepts of functions into more abstract realms.

Meanwhile, unaware of (or uninterested in) the controversy surrounding the vibrating string, Alexis Clairaut was calculating the complex motions of the moon.[20] Alexis Claude Clairaut (1713–1765) was a French mathematician who was elected as the youngest member of the Paris Academy in 1731 at the age of 19 for his work on the geometry of curves. He became close friends with Maupertuis, Voltaire, and du Chatelet, and he and Maupertuis studied together under Johann I Bernoulli in Basel.[21] Clairaut was welcomed into the Paris salons as a celebrity after his participation in a famous expedition to Lapland to measure a degree of longitude, confirming that the Earth's shape was an oblate spheroid, as predicted by Newton. However, Clairaut's chief claim to fame came in 1759 when he correctly predicted the return of Halley's comet based on improved calculation techniques he had developed to calculate the motion of the moon. This mundane task—calculating the motion of the moon—had great importance for the navigation of ships that had no reliable way to determine their longitude at sea. Despite the seemingly simple problem of the single body of the moon orbiting the Earth, the effects of the sun on the Earth–moon system, as well as the slight non-sphericity of both the Earth and the moon, made this an exceedingly difficult problem.

To perform calculations on the quasi-periodic orbital-mechanics problem of the moon's orbit, Clairaut used series of sinusoids to capture deviations from simple elliptical motions. This work made him aware that the integral over two sinusoids of different periods vanishes, while the integral of a sinusoid multiplied by itself produces a constant. This statement is what we would today call the general orthogonality of harmonic functions, but for Clairaut it simply provided a convenient shortcut that eliminated the need to calculate many integrals. These early, seemingly unrelated findings on the addition of sinusoids would be formalized many years later by Joseph Fourier into the field of mathematics called Fourier analysis.

Fourier's political astuteness had landed him one of the top intellectual positions of Napoleon's Egyptian campaign when he became the head of L'Institute d'Egypte in French-occupied Egypt and later the editor of the incomparable Description de l'Egypte. His prominent position came in part from his knack for playing the middle, yet he had barely survived the Terror during the French Revolution. At one point he was arrested by Robespierre and was bound for the guillotine when Robespierre himself was guillotined, and Fourier was set free. A year later, when the horrors of the Terror were rejected, Fourier was ironically arrested and imprisoned under false suspicion for being a supporter of Robespierre. Fortunately, Fourier's supporters at the Ecole Polytechnique worked for his release, and he escaped the guillotine once again. Fourier was settling back into the life of lecturer at the school when the head of the Polytechnique, Monge, picked him for Napoleon's Egyptian expedition, truncating his academic career once again.

After his long-delayed return to France in 1801, Fourier hoped to continue his teaching duties at L'Ecole Polytechnique, but Napoleon recognized his talent as an administrator and appointed him as the prefect of the Department of Isere centered at Grenoble. Despite his reluctance, he was good at it and made important improvements in local infrastructure, for which Napoleon rewarded him with a barony in 1808. However, when Napoleon was exiled in 1814 and passed through Grenoble on his way to Elba, Fourier was astute enough to be absent. Yet when Napoleon returned, there was no avoiding him, and this time Napoleon made him a count and promoted him to the prefect of Lyons. Napoleon's Hundred Days were brief, ending at the Battle of Waterloo in 1815, which also ended Fourier's noble title and his administrative position, putting him in immediate disfavor with the returned Bourbon monarchy.

In the midst of these political ups and downs, Fourier somehow found time to pursue studies in math and science. While at Grenoble, he developed a theoretical approach to the physics of heat, deriving the partial differential equation for heat conduction and solving it for given boundary conditions. As Daniel Bernoulli had proposed for the vibrating string, Fourier showed that any function on a given domain could be expressed as an infinite trigonometric series. Most significantly, Fourier derived a direct way to calculate the coefficients of the series, today called Fourier coefficients, with a simple integral method that Euler and Daniel Bernoulli had previously sought but had not found. He presented the work in 1807 to the Academy where his supporters Monge,

Laplace, and Lacroix approved its acceptance. But an elderly Lagrange rejected the work and would not budge in his decision. Lagrange had been on the losing side in the nasty battle over the solutions to D'Alembert's wave equation for a heavy string, and now years later he was still entrenched. He did not believe that an infinite trigonometric series could describe any function on an interval that included boundaries. He criticized Fourier's attempted proof of the principle, and the paper was rejected.

Fourier regrouped and sought to improve his manuscript, extending the boundaries of the domain to infinity and replacing his trigonometric series with an integral expression that is today called the Fourier transform. He submitted his extended method to a prize competition in 1810 on heat diffusion. His entry was selected as the best submission and he won the competition, but Lagrange still held sway, and the prize committee continued to criticize Fourier's method of proof. Lagrange's criticisms, though overzealous, were partially founded, for mathematics of the day did not yet possess the rigor to complete the proof. The subtleties went deeper than anyone at the time could have imagined, and it would take the genius of Bernhard Riemann decades later to finally establish the convergence properties of Fourier series and integrals. Nonetheless, Fourier finally published his theory in 1822 in his book *Théorie analytique de la chaleur.*[22] Fourier analysis is a theory of linear superposition, and although it was originally used in the solution of heat problems, it provides the fundamental mathematical underpinning of the superposition of waves and interference. Yet its delayed publication withheld its mathematical power from the very person who was seeking to establish the principle of superposition and interference of light waves—Fresnel. Nonetheless, Fresnel was able to construct a form of Fourier analysis on his own that gave him the foundation on which to build toward the ultimate victory of the wave nature of light over Newton's corpuscles.

An Academy Award

In early 1817, the battle in Paris between waves and particles was being contested as bitterly as ever, with Laplace and Poisson pitted against a nearly solitary Arago. In a gambit to put a stop to the foolishness of the undulationists, Laplace and Poisson arranged for the biannual prize of the *Académie des Sciences* to be given for the best solution to the problem of diffraction, fully expecting to receive a definitive account of the ray theory from one or more of their carefully coached students. The prize was announced in the spring of 1817 with a deadline of August 1818 and to be awarded in 1819. The wording of the prize problem was entirely in the context of the corpuscular theory of light, leaving little room for a contribution from a wave point of view. Nonetheless, Arago alerted Fresnel to the opportunity, expressing his desire that Fresnel submit his latest work.

By late fall, Fresnel recognized a central feature of the problem that allowed him to use the principle of superposition to add two waves that were 90 degrees out of phase.[23] If the amplitudes of the two waves were A and B, then the resultant intensity was simply the sum of squares $I = A^2 + B^2$, just like Pythagoras' theorem for a right triangle. The two wave amplitudes for this special case add together as if they were vectors. In a direct generalization of this picture, Fresnel realized that two waves that were *not* "orthogonal," with some phase difference that was *not* 90 degrees,

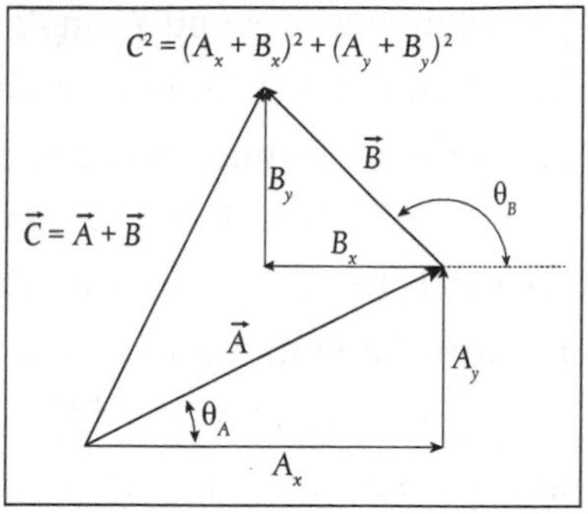

Figure 2.5 Fresnel's algebraic addition of two arbitrary waves as if they were vectors. The resultant detected intensity is $I = |C|^2$.

could be broken into two constituent waves that *did* have the required 90-degree phase difference. This is like starting with the hypotenuse and finding the two sides of the right triangle. The x-components of each vector are added and squared, then added to the square of the sum of y-components, as shown in Fig. 2.5. The resultant intensity detected in an experiment was then simply the square of the amplitude of the resultant vector.

Here we see the central step in the development of modern diffraction theory. It is based on vector addition, even though waves were certainly not thought of as vectors at that time. Multiple waves are simple sinusoidal functions with some phase relationship among them. Fresnel's insight, his flash of genius, was recognizing that the addition law for waves was the same as the addition law for vectors. Fresnel had independently discovered the "orthogonality" of sinusoidal functions first noted by Clairaut and contemporaneously being formalized (unknown to Fresnel) by Fourier for his later 1822 publication. It must have been with some excitement and pride when, in January 1818, Fresnel wrote up and submitted his principle of wave addition as a supplement to his previous papers. This was the pivotal moment in his career, propelling him beyond Young's qualitative approach, and arming him with a powerful new mathematical weapon with which to attack the most difficult diffraction problems with absolute rigor.

Within only a few months, Fresnel developed an integral approach to wave diffraction that divided the free aperture of a problem into infinitesimal areas that each were a source of a secondary wave. To find the diffraction caused by light illuminating any obstacle, one integrated (summed up) over all the infinitesimal areas, even if they extended to infinity. These integrals are today called Fresnel integrals. He submitted his initial entry to the Académie prize on April 20, 1818. The submission was anonymous, signed with the epigraph Natura simplex et fecunda (Nature simple and fertile) that was also written in a sealed envelope with his name and deposited with the secretary of the academy. The note was followed at the end of July with a complete manuscript on his new method and results. Fresnel had worked through 25 cases of diffraction and compared three of them with careful measurements of diffraction from a half plane, a narrow aperture, and an opaque strip. The agreement between theory and experiment was nearly perfect.

The prize committee was first composed entirely by the emissionists Laplace, Biot, and Poisson, but Arago complained and campaigned until he was added to the committee along with Guy-Lussac, who was neutral in the emissionist/undulationist debate. Arago was even allowed to become the chairman, since the emissionists had a secure majority. Despite the high profile of the prize, only two submissions were received. They were both anonymous, but the memoir by Fresnel was easily recognized. The other submission was emissionist, so the committee had an apparent choice to make between the competing theories. As the committee members began analyzing the submitted theories, the sophistication of Fresnel's entry clearly placed it above the other, so the committee really had only one choice: to award the prize to Fresnel or award no prize at all.

Although Poisson was antagonistic toward Arago because of their earlier political conflicts, he recognized the mathematical merit of Fresnel's work. The Fresnel integrals were a powerful new tool that enabled virtually any simple diffraction problem to be solved, and Fresnel had included many examples in his entry. But one problem that Fresnel had not considered was the diffraction of an opaque circle. Poisson tackled this problem himself, using Fresnel's new method. It must have been with some glee that Poisson reported to the committee an absurdity he had discovered using Fresnel's technique. According to his calculations, a small bright spot should occur at the very center of the shadow of the disk regardless of how large the circle was or how close the observation screen was placed. This nonsensical prediction, arrived at using Fresnel's own technique, was now turned against him and seemed to decide against Fresnel receiving the prize, but Arago was not swayed.

Over the three years of his association with Fresnel, Arago had witnessed time and again Fresnel's keen insight into physical phenomena, his ingenious and accurate experimental measurements, and his mathematical prowess. Arago must have been confident that Fresnel had discovered the fundamental cause of diffraction, based on the wave nature of light, and that Poisson's spot must certainly exist. The spot would not easily be seen, because it was clear from Fresnel's theory that the circle needed to be smooth to within a fraction of the wavelength of light, which was no easy feat. He enlisted the help of the instrument maker François Soleil who fashioned a precise metal circle with a diameter of about 2 mm, attached by wax to a flat glass plate. In his later memoirs Fresnel recounted the famous experiment, describing how precisely the disk edge needed to be fashioned, and the need to have exceedingly flat glass.[24] Indeed, the care that is necessary to see the spot speaks to the highly refined optical craftsmanship in the workshops of Paris.

The details of when Arago observed the tiny bright spot in the shadow of the circle and who was present are unknown. It is not likely that the experiment would have been performed for the first time with the committee present, because all experimentalists know that nothing works on the first try. Arago would have worked with painstaking care to get just the right alignment, and to get all the best conditions to see the spot. Only after he had arrived at a robust operating point, where the spot continued to be observable as conditions were altered slightly, would he have risked bringing the committee in to observe the results. The final demonstration was probably carefully choreographed, almost certainly done with dramatic flair for maximum impact on the emissionists Laplace and Biot as well as Poisson, who had the honor of having predicted this unusual phenomenon. The phenomenon of the Spot of Arago is shown in Figure. 2.6.

The report of the committee in favor of Fresnel was unanimously delivered to the *Académie des Sciences* on March 15, 1819. The predictive power of Fresnel's new method was uncontestable, and direct demonstration trumped any theoretical arguments the emissionists could make against it. The unanimous report spoke to the reality of what Fresnel's approach could predict, and the emissionists could put their names on the report with full confidence that the results were correct. However, accurate

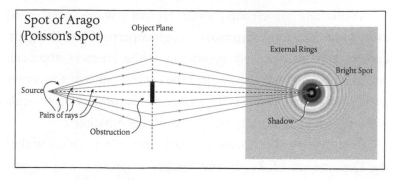

Figure 2.6 Spot of Arago (Poisson's spot). Pairs of wave paths, scattered from the object plane according to the Huygens–Fresnel principle, add constructively to form a bright spot in the center of the shadow of the opaque disk.

mathematical machinery was not necessarily the same thing as proof of the principle, and emissionists were free to use Fresnel's method without adopting the wave principles behind it. They would argue that the mathematics correctly captured the behavior of deflected rays with as yet undiscovered causes for the bright and dark fringes. They used Fresnel's techniques without ascribing physical reality to it and without embracing Fresnel's principle.

That principle today is called the "Huygens–Fresnel principle," attaching both names to it. This duality is deserved because of the similarity in appearances between what Huygens and Fresnel proposed and because of the essential new interpretation provided by Fresnel. Although both Huygens and Fresnel drew figures of outgoing secondary circular waves from a point on a primary wave, Huygens had conceptualized light as a succession of discontinuous shock waves, while for Fresnel they were continuous sinusoidal undulations. Huygens' shock waves only overlapped at the discrete wavefronts, while Fresnel's waves overlapped everywhere and were summed through superposition. Huygens' construction was entirely geometric, while Fresnel's approach was analytic. However, both held the central concept that every point on a primary wave is the source of a secondary wave—this is the source of the conceptual power of this principle.

After winning the *Académie* award, Fresnel's position within the Department of Roads and Bridges improved, and he was reassigned, with Arago's help, to the intellectually stimulating problem of improving the brightness of lighthouse lights. This was a theory and design project far different from overseeing uneducated workers on roads. Fresnel took to the new position with energy and quickly developed his ideas for the Fresnel lens that would replace old-fashioned mirror reflectors in lighthouses around the world. He also had time to pursue additional problems in the theory of light waves, leading him to discover important polarization-dependent effects, such as the Fresnel reflection and transmission coefficients that every undergraduate physics major learns today. He was unfortunately consumed by tuberculosis and died at the early age of 39 in 1827. Thomas Young, in a fitting tribute and acknowledgment of what Fresnel had achieved, translated Fresnel's *De La Lumiere* into English, which was published in installments between 1827 and Young's own death in 1829.

By 1830 most physicists had adopted the wave nature of light. The undulationists had "won" the fight against the emissionists, and physics had moved on, but hold-outs like David Brewster in England and even Poisson in France, still irked Arago. He believed that a final experiment was needed to put the nail in the coffin. Although he did not succeed in doing the experiment himself, he mentored those who did. That experiment, so important in Arago's mind, actually had little to no effect on the epistemology of light—but it would launch the very practical field of optical interferometry.

3

At Light Speed

The Birth of Interferometry

Arago had more ideas by himself than a full generation.[1]
Léon Foucault (1853)

O pen your eyes. The world goes from dark to light as reality presents itself to you in a flash of awareness. Distant objects touch your eyes as light rays strike your retinas and launch a cascade of neural computing that constructs your perceived reality in your mind. It would seem like magic if you weren't so used to it. Light is one of the most powerful manifestations of the forces of physics because it tells us about our reality. Light from the early years of our universe fills space with a cosmic background radiation that tells us of the Big Bang. Light from distant galaxies tells us of the subsequent history of an expanding universe. Light observed from the rotating belts of stars in nearby galaxies tells us of the existence of dark matter, and light from the eclipses of the moons of Jupiter told us (through Ole Rømer and Christiaan Huygens) that light had a finite speed.

Yet for all its power to advance science and its immediacy in our daily lives, light has presented an enigma to virtually every generation of scientists who have tried to grapple with its essence.

Interference. David D. Nolte, Oxford University Press. © David D. Nolte (2023).
DOI: 10.1093/oso/9780192869760.003.0003

Whether it was Descartes, Huygens, Newton, Euler, Laplace, Arago, or Fresnel—they all battled with light's ambiguity. No sooner had the wave nature of light been accepted, when new controversies sprang up about what medium such waves travelled through. Fresnel invented the first luminiferous ether to support light, but it paradoxically needed to be partially crystalline to support transverse polarization while being partially entrained inside moving matter like water. This launched another long list of scientists grappling with light. Whether it was Green, Stokes, Maxwell, Lorentz, Poincare, Michelson, or even Einstein—the nature of light refused to be corralled one way or another. Indeed, long after Fresnel was dead, Arago continued to wonder and worry, dreaming always of the "final" experiment to answer all questions. He was fortunate to live long enough to believe that he had seen the ambiguity end. And although he was too old to do the experiment himself, he saw the triumph of an apparatus that he had built with Fresnel in their early days at the Paris Observatory. That apparatus was the precursor to a family of later optical machines that ultimately would tame light—optical interferometers.

The Problem with Snell's Law

St. Michael's Church in the West Riding of Yorkshire, England, is a solid structure dating from the 1400s, constructed of yellow millstone with an imposing square tower looming over the church yard. The rector of St. Michael's Church from 1767 to 1793 was an unimposing clergyman, short and fat, who made little impression on those who met him casually. Yet for those who knew him better, he had a mind that raced far ahead of his time in fields of natural philosophy. John Michell (1724–1793) was educated for the church at Cambridge Queen's College where he later taught courses ranging from theology and geometry to Hebrew and Greek before being appointed to St. Michael's in the small town of Thornhill. Apparently, this move freed up his time and freed up his thoughts considerably, because he began to write the most astonishing papers, speculating on matters of light and electricity and magnetism that brought eminent visitors like Joseph Priestley, Henry Cavendish, and even Benjamin Franklin to his small-town parish home.

Michell was the first to recognize that the effects of earthquakes traveled outward from the source as seismic waves, and that they could be used to interrogate the nature of the rupture. He also invented a balance system that could measure local variations in the mass of the Earth. Due to these efforts, he is considered to be the father of the modern field of geophysics. Yet his interests were surprisingly broad. In the field of materials science, he devised an ingenious way to construct magnets that could be manufactured with good reproducibility rather than relying on the random properties of lodestones. In cosmology he recognized that binary stars were mutually bound by gravity, and by using statistical reasoning he argued that they should have a high abundance in the heavens. In fact, as many as 85 percent of all stars are in binary systems.

John Michell is probably best known for a paper he published in 1784 with the excessively long title "On the Means of discovering the Distance, Magnitude, etc. of the fixed Stars, in consequence of the diminution of the Velocity of their Light, in case such a diminution should be found to take place in any of them, and such other Data should be procured from Observations, as would be farther necessary for that Purpose."[2] In this paper he proposed that there could be gravitating bodies of such mass and size that they would prevent the escape of light from their surfaces, making them appear as dark stars. Being dark, they also would be invisible, yet he suggested their presence could be ascertained through binary systems in which a bright star orbited the dark star. To a modern reader, Michell's dark star is recognized as a black hole which often *are* detected through their effects on a stellar companion. What is not as well known about this remarkable paper by Michell is that, without knowing it, he opened a crack in Newton's formerly unassailable theory of corpuscles of light.

Michell knew from his astronomical studies that stars move with a range of speeds. According to Newton's theory of light corpuscles, stars moving toward the Earth should emit light particles that traveled faster, while stars moving away from the Earth should emit light that traveled slower. This was like shooting a bullet from a gun on a speeding train, adding velocity to velocity to give the bullet faster or slower speeds relative to the stationary railroad embankment. This part of Michell's argument would have been clear and obvious to any student of Newton's physics. Then he took the next brilliant step.

Michell was well familiar with Snell's law on the deflection of a ray of light when it enters a transparent medium. Snell's law was named after the Dutch mathematician Willebrord Snel van Royen who described the effect using a law of sines in 1621, although it was Descartes who first published the law in 1637. Michell recognized that Snell's law for the refraction of light by transparent materials, like glass, depended on the speed of light outside relative to inside the glass. According to Newton's theory, particles of light travel faster in denser materials. This would be why light deflects toward the normal if it enters a denser medium. The smaller angle inside the denser material keeps the transverse speed of the light particle the same. Yet Michell pointed out that if light particles were traveling faster or slower, depending on the speed of their emitting body, then the angle they make after entering a glass surface, like a prism, should differ slightly, because Snell's law is a function of the different speeds of light relative to the Earth-bound prism. Michell calculated the size of the effect and concluded that differences in the speed of light as small as one part in a thousand might be measurable. This is about 3×10^5 m per second, which is close to the speed of the Earth as it orbits the sun. Therefore, Michell's speculations were not merely philosophical but were actual experiments that might be tried.

In 1805, when François Arago was merely 19 years old, and yet to be sent on his epic journey to Spain to measure the meridian, he was appointed as the secretary-librarian of the Paris Observatory. He began testing methods to measure the speed of light, and he considered using Michell's method of variable refraction. However, rather than attempting to find if light emitted from moving stars had differing speeds, as Michell proposed, he decided to test the effect of the Earth's motion relative to light coming from a given star or set of stars. He observed the positions of stars through a fixed prism at different times of the year as the Earth's relative motion changed on its yearlong journey around the sun. To within 5 arcseconds he observed no change in the refractive angle or the positions of the stars.[3] He presented his findings at the Institute on December 10, 1805, but further measurements were postponed when he and Biot were sent on their mission to Spain the next year.

The problem of the speed of light must have stayed on Arago's mind, possibly helping him wile away the time in jail in Spain, because upon his return to France he took up the experiments again. This time he used a prism with a larger angle to eliminate dispersion while magnifying the effect.[4] He placed the prism so that it covered half of the objective of the telescope. Then all one needed to do was observe a star through the prism, then shift the eye slightly to observe it through the objective alone and look for a shift in position. He placed a graduated circle within the same field of view, allowing very precise measurements of the apparent position of stars down to a few seconds of arc. The motion of the Earth under these conditions would produce a displacement, depending on the time of year, that varied by as much as 6 seconds of arc. The effect should have been observable given the accuracy of the method, yet the experiments produced null results.

Arago's measurements on the refraction of starlight created a serious problem for the Newtonians who believed firmly that light was composed of particles. To save the emissionist theory, ad hoc ideas were suggested. For instance, Laplace, in his latest edition of his *System du Monde*[5] in 1813, questioned whether the eye could detect light of different speeds. Arago himself was puzzled, yet he was confident in his experimental results. He knew that if Newtonian theory was correct, then his measurements would have found the effect of the Earth's motion. Thus, when Merrimé Fresnel approached Arago with the ideas of his nephew Augustus about the wave nature of light, Arago was primed and ready to overthrow the emissionists—especially since he was by then politically opposed to Laplace.

The Dawn of Interferometry

Sometime during the two years from 1816 to 1818, when Fresnel and Arago worked together at the Paris Observatory, Arago realized that the separated paths in Young's double-slit experiment provided an opportunity to have one path travel through air while the other could pass through a transparent material. For instance, a thin sheet of mica could be placed over one of the slits while the other slit remained clear. When Arago and Fresnel performed this experiment, they noticed that the fringe pattern shifted by a non-negligible amount. Based on their wave theory of light, the shift represented a delay in the arrival of a wave front at the observation plane caused by the refractive index of the mica slowing the speed of light. It can be argued that this experiment, by Arago and Fresnel, was the first interferometer, because it was the first time interference was used to measure a material property.

Although this demonstration of interferometry was crude, it is common with many discoveries that, once the new phenomenon is observed, subsequent improvements move rapidly. Within a year of this simple double-slit experiment, Arago had constructed a much more sophisticated interferometer that could measure refractive index changes across long path lengths. Late in his life, in 1852 just before his death, Arago wrote his memoir, describing the interferometric experiment he performed with Fresnel in 1818 to measure the refractive index of moist air relative to dry air. This was an important question for astronomy because changing atmospheric conditions place fundamental limitations on astronomical observations of slight shifts in stellar positions. For instance, starlight might be affected by the change in air moisture content as a function of altitude by refracting light, causing systematic shifts in their positions. As an astronomer, Arago was well aware of the possibility of the effect and sought to answer whether it could affect astronomical observations.

Arago's interferometer design was ingenious. The key problem was to get two independent light beams, originating from the same small source, to propagate parallel to each other with a wide-enough separation that one beam could travel through one material while the other path traveled through a different material. The double-slit configuration is not appropriate for this because the beams spread out widely when they diffract from the slits. Fortunately, Arago knew that a lens can take light from a point source and turn it into a collimated beam that can travel over many meters without expanding. This was achieved simply by placing the point source at the front focal point of the lens. Masking off the light after the lens created two parallel pencil beams. One pencil of light passed through a copper tube filled with moist air, while the other traveled outside the tube in dry air. To recombine the two paths to create interference fringes, the process was reversed, using a lens to focus the light back to a small spot where fringes could be observed by eye.

The final apparatus, shown in Figure. 3.1, had two side-by-side copper tubes a meter long that were capped with equal-thickness optical-quality plates. Moist air was introduced into, or pumped out of, one of the tubes. A lens collimated the light from a small pinhole, and an opaque mask allowed each beam to enter into the copper tubes, propagating without reflecting from the tube walls. After exiting through the glass plates, a lens focused the two beams down into one. When the path lengths were identical, fringes formed at the output, and as moisture was pumped out of

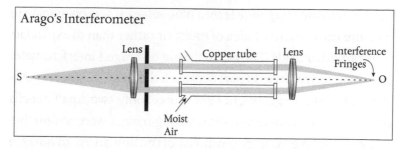

Figure 3.1 Arago's interferometer from around 1818. The point source at S is collimated by the lens and passes through two apertures. The beams traverse the 1 m-long copper tubes and are recombined by the second lens. Moist air could be added to or removed from one of the copper tubes. The second tube serves as the reference arm.

one of the tubes, the fringes shifted perceptibly. Light propagating through a meter of saturated moist air suffered a fringe displacement of about one and a quarter wavelength,[6] representing a refractive index of moist air that differed from dry air by about one part in a million.

This exquisite measurement was so clever and so sensitive that it created a stir in the scientific community. Laplace, who was no great friend or supporter of Arago, mentioned it several years later in his 1824 edition of his *System du Monde*, writing that "M. Arago, using a means as accurate as ingenious, showed … the influence of the humidity of air."[7] The effect was so small that nothing other than interferometry could have measured it. It was also so small that it removed the question of whether the humidity of the atmosphere created a problem for stellar positions—it did not.

Despite the obvious usefulness of Arago's interferometer from our perspective today, it was a one-off experiment answering a very specific question at that time, and then it was abandoned. Even 30 years later, when Arago revisited the experiment with the help of Armand Hippolyte Fizeau (1819–1896), it was once again to measure the refractive index of moist air rather than to expand its uses into other fields. The gap in the use of Arago's interferometer for nearly 30 years was in part because it did not use light very efficiently. From a single bright light source, only two small pencils of light were extracted, and the resulting fringes were so dim that they required the extreme sensitivity of the human eye to observe the fringes in a very small region where the beams overlapped. Arago's interferometer, although brighter than Young's original double-slit experiment, still rejected the vast majority of the light rays.

This disadvantage was later removed by Fresnel who conceived of an optical configuration using two slightly angled mirrors that captured all the light rays to produce bright fringes. The fringes could be observed over a large area as opposed to being restricted to the tiny spot of fringes that was produced by Arago's interferometer. Fresnel's mirror in Figure. 3.2 consisted of two

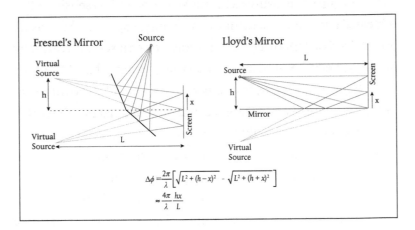

$$\Delta\phi = \frac{2\pi}{\lambda}\left[\sqrt{L^2 + (h-x)^2} - \sqrt{L^2 + (h+x)^2}\right]$$

$$\approx \frac{4\pi}{\lambda}\frac{hx}{L}$$

Figure 3.2 Fresnel's mirror and Lloyd's mirror. Interference fringes appear on the screen as the phase varies linearly with height.

mirrors placed side-by-side but with a small angular deflection of one relative to the other. A light source illuminated the mirrors at an oblique angle, reflecting the rays onto a screen. Because the mirrors were tipped at slightly different angles, they produced two virtual images of the light source that were slightly displaced from each other, filling space with overlapping wavefronts that produced high-contrast fringes over a large area of the screen.

Fresnel's mirror was not used as a refractometer, at least not by Fresnel, but it did provide a bright, dramatic, and easily viewed demonstration of the interference of light. Fifteen years later the Irish physicist Humphrey Lloyd (1800–1881) created a version that was even easier to operate because it used only a single mirror and a screen.[8] These bright wide-field interferometers were the predecessors of holographic gratings that would become common after the invention of the laser in 1960. They also played a particularly important role in the first demonstration of the temporal coherence of light performed by a famous duo of French optical physicists.

Fizeau and Foucault

The French physicist Armand Hippolyte Louis Fizeau (1819–1896) was born in Paris into a wealthy family as the son of a successful doctor. He was set on a path to follow his father into medicine but was diverted when he took a course in optics. Armand was not considered the brightest of the Fizeau children, and he attended the College de France while his brother attended the prestigious Ecole Polytechnique. His brother took detailed notes in his own optics course that he then passed to Armand who studied them assiduously. What started out as an intense hobby in physical optics turned into a vocation when he began studying at the Paris Observatory under the tutelage of François Arago, who recognized an untapped talent in this largely self-taught student.

Arago by the late 1830s had become the premier French expert on optics and optical phenomena and was its chief proponent and archivist. He wrote and lectured prolifically, occasionally suggesting new experiments, although his own days of experimental work were long passed. Following the July Revolution of 1830 (the insurrection that was featured in Victor Hugo's famous book *Les Misérables*) Arago had become active in politics. Over the following two decades, Arago was elected and re-elected as a deputy for his native department (Pyrénées-Orientales), and later for Paris, where he sat on the left in the *Chambre des Deputies*. He was a proponent of educational reform and freedom of the press as well as the application of scientific knowledge to technological progress, particularly canals, steam engines, railroads, the electric telegraph, and photography.

On August 19, 1839 Arago gave a public lecture on a new invention by L. -J. -M. Daguerre that allowed the image formed in a camera obscura to be captured permanently on a light-sensitive plate. The daguerreotype process, named after the inventor, became a sensation that was taken up by amateur and professional scientists. Fizeau was fascinated by the daguerreotype and began experimenting with ways to improve the contrast of the prints. Fizeau met a kindred spirit in Jean Bernard Léon Foucault (1819–1868) who also had been a weak student, but who had tremendous experimental skills and a curious mind. Like Fizeau, Foucault had initially thought to become a doctor and showed particular promise as a future surgeon, but he was repulsed by the gore and suffering that accompanied that profession. Fortunately, his abilities were noticed by a professor of microscopy who took him on as a laboratory assistant, drawing Foucault into the study of light and optics.

Fizeau and Foucault formed a close friendship and forged an informal team to study and improve early photography. Beginning in 1843 they worked for two years on photographic processes that culminated on April 2, 1845, in the first scientific photograph of the surface of the sun that was able to clearly delineate the images of sun spots.[9] By this time, they also had been experimenting with interference effects, initially following Young and Fresnel, but then forging beyond with new approaches that struck to the core of the difficult physics of coherence. In November 1845 Fizeau and Foucault performed one of the most elegant optical experiments of the nineteenth century. Today the experiment would be called spectral interferometry, but at that time it had no precedent.

A persistent problem with interference effects had been the extreme sensitivity of the effects to the different path lengths taken by separate interfering waves. Under the best conditions, with the brightest sources, it was only possible to detect about half-a-dozen fringes to either side of the main interference peak. The cause of this limitation was known to be the broadband nature of white light that was composed of many different wavelengths. When the optical path difference between two beams of light was zero, then all the wavelengths aligned in perfect constructive interference. However, when the optical path difference increased to half-a-dozen wavelengths, then the longer and shorter wavelengths no longer provided perfect interference, and the fringe visibility vanished. There had been

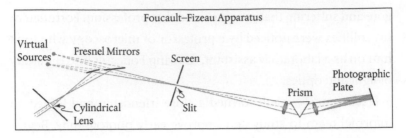

Figure 3.3 Spectral interferometry of Fizeau and Foucault (1845).

attempts to pass white light through colored filters to make it more monochromatic, but too much intensity was lost in the process.

Fizeau and Foucault hit upon a brilliant solution to this problem by using the Fresnel mirror to produce relative path differences while maintaining high intensity, then dispersing the interfering beams through a prism to create a spectrum.[10] By using a set of lenses, they recombined the spectra on an observation plane that held a photographic plate, as shown in Figure. 3.3. The result was a bright spectrum spanned by thousands of fringes.[11] Path differences as large as 7,000 wavelengths (several millimeters) could still produce interference fringes.[12] This experiment was the first demonstration of the classic trade-off between spectral bandwidth and temporal coherence length. Not stopping there, Fizeau and Foucault extended their measurements on the long-wavelength side of the spectrum by using tiny thermometers to measure the first interference effects for infrared light.[13] They also placed polarizing crystals in the system and demonstrated the first interference fringes represented by polarization rather than by intensity.

These experiments by Fizeau and Foucault went far beyond the work of Fresnel who, in his own time, had gone far beyond the work of Young. Fresnel had replaced Young's double slit with alternative geometric configurations that firmly illustrated the wave nature of light, but which did not delve into questions of coherence, other than recognizing that spatial coherence was required and could be created through the use of pinholes. The work by Fizeau and Foucault launched the field of coherent optics, which today is a vast field on which vast tomes have been written.[14] Coherent optics is a workhorse of modern optical techniques and applications. For instance, spectral interferometry is used for the characterization of femtosecond laser pulses, and it makes it possible to image inside translucent media in a format known as spectral-domain optical coherence tomography. Arago, who had begun as an emissionist, must have beheld all this with some sense of wonder, witnessing within his lifetime the development of interference from the simple slits of Young, through the improvements by himself and Fresnel, to the exquisite demonstrations by Fizeau and Foucault.

Twilight in Paris

Some years before Fizeau met Foucault, Arago had attended a meeting of the British Association of Science where he saw a presentation by Wheatstone that would prove oddly fatal to the fruitful collaboration and friendship between Fizeau and Foucault. Charles Wheatstone (1802–1875) was a British physicist who became famous during the Victorian age for his contributions to the telegraph and especially for what is today known as the Wheatstone bridge. In 1838, before the teenage Victoria ascended the throne, he built a high-speed rotating mirror device for the ostensible purpose to measure the speed of electricity propagating down a long wire. As soon as Arago saw it, he realized that it could be used to answer the question that had obsessed him for his entire career: the question of the wave-like versus particle-like nature of light. By 1838 this question had been firmly answered in favor of the undulationists, but the earlier battles with Biot and Poisson must have left a mark, because Arago still yearned to make the decisive experiment to answer the question beyond a shadow of a doubt.

Based on the old battle between the emissionists and the undulationists, it was expected that if light were a wave, then it should travel faster in air, but if light were a particle, it should travel faster in water. The only measurements on the speed of light up to that time were astronomical, for no one had devised a method to measure the speed of light terrestrially, and certainly not in a laboratory. When Arago returned to France in the summer of 1838, he gave a public lecture in which he proposed to use the rotating mirror to make a direct comparison of the speed of light through air versus water. At that time, such a public declaration represented the claim of priority for the idea, and Arago had every intention of carrying it through. Not held hostage entirely to his political life, Arago built a prototype for his proposed apparatus and attempted to perform the experiment in 1844, but the results were ambiguous.

By this time Fizeau had been working with Arago for several years, and it was the same year that he and Foucault began working on daguerreotypes. Therefore, ideas about experiments to measure the speed of light directly in a terrestrial setting were never far from the minds of all three scientists. Unfortunately, the spinning-mirror experiment proposed by Arago had an optical configuration that made calibration very difficult, putting the experiment on hold.

Several years later, in February 1848, Fizeau had a flash of insight into how to measure the speed of light directly without the spinning mirror. This was not the experiment that Arago was most interested in—Arago wanted to compare the speed of light traveling in air relative to water—but it would represent the first terrestrial measurement of the speed of light in air. The idea was surprisingly simple, using a spinning toothed wheel instead of a spinning mirror. In Fizeau's design, illustrated in Figure. 3.4, the teeth of a spinning wheel would periodically block a beam of light as it reflects off a distant mirror and returns along the same path to the observer. Because light takes a finite time to propagate, if the wheel is spinning fast enough, an open notch for the outgoing beam will have shifted sideways so that an opaque tooth now blocks the return beam. At higher speeds, the returning light would pass through the following notch and becomes visible again, and at even higher speeds the light would be blocked yet again by the next tooth. By measuring the speeds

Figure 3.4 Fizeau's experiment to measure the speed of light.
Lequeux, *Arago*, p. 113.

at which the returning light becomes visible or is blocked, given the distance to the reflecting mirror, one could measure directly the speed of light in air. Because the idea had been Fizeau's alone, it was understood that he would work on the experiment without involving Foucault. This was a collegial decision that should not have sparked jealousy, except for an unfortunate twist.

In January 1849, just before he began working on his speed-of-light experiment, Fizeau was elected to be member of the prestigious *Société philomathique de Paris* in recognition of the work he had done with Foucault on the daguerreotype, as well as on the uses of photography in telescopes for astronomy and for the elegant spectral interferometry experiment, also with Foucault. Foucault was thrilled for his friend and collaborator, in part because he was next in line for election to the *Société* later that summer. It had been the custom for the second-place candidate to be selected in the following election. However, the next election committee did not follow this tradition and passed over Foucault. This slight may have been because Foucault's main occupation, when he wasn't in the laboratory, was as a science reporter for the French magazine *Journal des débats* where he was called on to give his opinions on the latest discoveries. He was sometimes less than complimentary, penning critical reviews that at times called experiments uninspired or pedestrian. These reviews may have built up resentment against him, and some of his targets would have been members of the *Société*. Regardless of the reason, the snub by the *Société* fell hard on Foucault who withdrew in disgust, and he may have fallen into a depression, for he disappeared from sight for nearly six months. It was during Foucault's absence that Fizeau completed his speed-of-light measurement.

Through the spring and early summer of 1849, Fizeau set up his experiment to cast a beam of light from the roof top of his father's house in Suresnes on the west side of Paris directed to a telescope mounted in a telegraph station on the hillside of Montmartre where the focused beam was reflected by a mirror back through the telescope, retracing its path to the rim of the spinning toothed wheel, where the return light was detected by eye. The total distance traveled by the light beam was an astounding 17.266 km. The experiments were performed in the Paris twilight when it was just light enough to enable beams to be directed to their dimly visible targets, while just dark enough to enable the extremely faint reflected light to be detected by a dark-adapted eye. After months struggling with the difficult optical alignment, the experiment was successful. On July 23, 1849, Fizeau reported his findings to the *Académie des Sciences*. He had measured the speed of light in air to be 315,300 km/sec, close to the previous value obtained astronomically.[15]

Fizeau received great acclaim, recognized for accomplishing one of the most important yet difficult experiments in the physics of light. Four months later, Foucault was belatedly elected a member of the *Société philomathique* (Fizeau was on the selection committee) and he returned to his post at *Journal des débats* where his first article praised Fizeau, proclaiming "That he has sometimes been lucky, I concede, but it is the same luck that supplies rhyme to the poet and discoveries to men of genius."[16] It is not known whether Fizeau was pleased with being called "lucky," because the design and hard work he had put into the experiment had nothing to do with luck.

At the beginning of 1850 it would have been natural for Fizeau and Foucault to have turned their attention to the experiment that Arago considered to be the real point of the speed-of-light measurements—determining the relative speed of light in air versus water. The experimental plans had hit a wall because a direct measurement of speed, like that done by Fizeau, was not accurate enough to measure the difference over short laboratory distances. Arago's design of the spinning mirror also had a fundamental alignment problem associated with the fold mirror. Once again Fizeau had a flash of insight and understood that the problem was because the fold mirror was flat. But if it were a curved concentric mirror, then the misalignment could be zeroed out. Such an insight, removing a barrier that had stood in the way for more than a year, should have made the collaboration move ahead smoothly with the main experiment. However, as Alfred Cornu, the French physicist who was a protégé of Fizeau, lamented years later, the day Fizeau had the idea of the concentric mirror was the last day of the collaboration.[17] It started a steeple chase as Fizeau and Foucault each forged ahead on their own to make what they thought would be a historic measurement that would stand in the annals of science for all time.

The specifics of why the two scientists fell out have been lost to history. A flurry of heated letters filled with recriminations went back and forth between Fizeau and Foucault during the last days of the competition, but the original cause of the split is unknown, and neither Fizeau nor Foucault ever spoke of it later through their long careers. One possibility is that Foucault may have made a *faux pas* in perceived etiquette with respect to Arago. Unlike today, when any scientist can take inspiration from another and perform any experiment without reproach, in the polite days of the French *Académie* there was a tacit ownership of experimental ideas. When Arago outlined his experimental design to test the speed of light in air relative to water, he had staked his ownership. Even if Fizeau and Foucault helped him with the experiment, it would have been understood that the experiment was his and that he should have the first shot at performing it. However, by early 1850 Arago was losing his eyesight because of diabetes. It must have been a severe disappointment to him, not only because of the difficulties it caused him personally, but also because it prevented him from completing the central experiment that had obsessed him for nearly his entire career.

At just this delicate moment in his life, Arago received an unexpected visit from Foucault. Intending no disrespect, and indeed because he respected Arago so highly, Foucault felt bound by honor to inform Arago that he wished to move ahead to complete the experiment by himself. Arago must have been taken aback. Foucault did not seem to be asking permission but was merely *informing* Arago of his intentions. Arago would have understood that, because of his own failing eyesight, the experiment was no longer within his power to accomplish, so he grudgingly gave Foucault his approval to proceed.

What motivated Foucault to take this bold step? It is possible that he felt his own career was being eclipsed by Fizeau's recent success and fame for his measurement of the speed of light. Or he may have been bruised by having been overlooked initially for the election to the *Société philomathique* (although he was now belatedly a member). On the other hand, as a master instrument maker, he had conceived of a mechanical device that radically improved upon Wheatstone's spinning mirror, enabling it to spin at much higher rates with much greater stability. Just as Fizeau's novel idea of the toothed wheel had allowed him to perform his speed-of-light experiment without Foucault, perhaps Foucault felt that his new design of the spinning mirror gave him the authority to do the experiment without either Fizeau or Arago.

Whatever his motivations, Foucault implemented his new spinning-mirror system using a steam-powered turbine that spun the mirror at an astounding 200 rotations per second. Despite the small size of the mirror, the centrifugal force was so great that it caused the mercury amalgam (used at that time for mirrors) to separate from the glass and splatter across the experimental benchtop. This forced a stop to the experiments until Foucault devised a method for depositing a sturdy layer of silver on the glass mirror. (This method of "silvering" a mirror was to have great significance in the later development of interferometers by Albert Michelson and others.) With baffles placed around the experiment to protect the optical elements from oil and steam flying off the screaming turbine, Foucault got a first glimpse on February 17, 1850 of the success of his ideas, though he still had several months of hard work ahead.

Around this time, either because Foucault shared his progress with Fizeau, or because Arago had informed Fizeau of Foucault's strange visit, Fizeau realized that he was behind in the race. Arago had previously enlisted the help of his personal instrument maker to build Wheatstone's spinning-mirror system for his own planned experiment, so he now loaned the apparatus to Fizeau to help him compete against Foucault. However, Foucault's design was superior, enabling him to move faster and approach the ultimate goal. By April 22, 1850 Fizeau was in a panic. He sent Foucault a rambling long letter, scribbled in crayon, accusing him of great insensitivity and misconduct. Foucault's response was short and polite, written in clean ink penmanship, begging Fizeau to "calm himself" and not to take such offense.[18] Fizeau replied with yet another letter, incensed and clearly viewing himself as the injured party. On April 25 the two longtime friends and collaborators split, unable to achieve a rapprochement. They would never work together again.

Two days later, on April 27, Foucault's experiment was successful. His two light paths, one through air and another through a 3 meter-long pipe containing water with glass flanges at the ends, displayed different displacements at the observation port. One image was shifted slightly to the side of the other in just the right direction to finally and definitively put to rest the emissionist theory of light. Light was composed of waves! Foucault was so excited, and so awed by the historical importance of the experiment, that he had three individuals peer into the view port to confirm the results.

Somehow Arago must have heard the news, because two days later on April 29 he gave a strange speech at the *Académie* outlining the history of his idea for measuring the relative speeds of light in air and water. He reminded those present of his original experiments many years before with Fresnel on the first interferometric measurements of the refractive index of dry air relative to moist air, but how this had contained an ambiguity on relative speeds, motivating his idea for a direct experimental test. He mentioned Foucault's name, who was sitting in the audience, but he did not mention his results. The next day, in a short paragraph in *Journal des débats*, Foucault announced that he had succeeded in making the experiment. Then on May 6, 1850 he officially reported the results to the *Académie*. His mirror had achieved speeds up to 800 rotations per second, producing deviations of 300 microns (measured by observing the shift using a thin fiduciary ruling) between the two light beams passing through air relative to water.

Fizeau followed Foucault by reporting that he too had constructed an apparatus that performed as required—but that the lack of sun had prevented him from making the measurements. At the end of his presentation he made the excuse that he had been delayed in starting his experiment because he had waited for Arago to "authorize us to embark on a topic of research which belonged to him."[19] This barb was clearly aimed at Foucault and his perceived indiscretion. It took Fizeau seven more weeks before he could report on his own experimental findings, but by then he had lost both the priority and the fame.

The aftermath of the Fizeau and Foucault affair is mixed. The importance of the air-versus-water experiment that loomed so large in their minds is actually a side note in the history of science. By 1850 the emissionist theory of light was already dead and only Arago was left worrying about the "final" proof. Foucault certainly believed that this experiment would be his legacy, making him immortal, which is probably why he was willing to step over the line to achieve it. Yet from our twenty-first-century perspective it is Fizeau's original experiment on the speed of light that stands out. That is how he achieved his own immortality even though his later interferometric work may have been more influential for the ongoing process of science. For Foucault, on the other hand, the pendulum that he demonstrated a year later in the Panthéon in Paris cemented his name into history even as his experiment on air-versus-water has all but faded from memory.

Interferometry Comes of Age

Fizeau understood the deep complexities of the coherent properties of light better than anyone else living. For instance, as early as 1848 he proposed that light from a moving source would be detected at shifted wavelengths in a spectrometer.[20] This phenomenon is the Doppler effect, named after Christian Doppler who had proposed the effect six years earlier in a small meeting of the Bohemian Society of Science in 1842.[21] Doppler had thought that the effect would make stars change color, while Fizeau correctly pointed out that the shifts would be detected in the change of wavelength of narrow spectral lines. The effect of Doppler–Fizeau now underlies the most sensitive astronomical measurements of motions across the universe. Fizeau also proposed additional astronomical applications of interferometry, as we will explore in Chapter 6 on stellar interferometry. Yet the experiment that he is most famous for, after his measurement of the speed of light in air, was his measurement of the speed of light in moving water. This was not merely an incremental advance on his earlier experiment, because 50 years after Fizeau completed these experiments, his results would provide the central evidence that Einstein would use to abolish the ether entirely.

Fizeau's experiment to measure the speed of light in moving water is shown in Figure. 3.5 as a modified form of Arago's interferometer. The light source is partially reflected by a glass plate to pass through the lens located one focal distance away from the source and is split by an aperture mask into two pencils of light. A pair of rectangular prisms separates the pencils so that they can pass through the flow apparatus composed of two tubes. Water is introduced into one side and expelled on the other, creating two

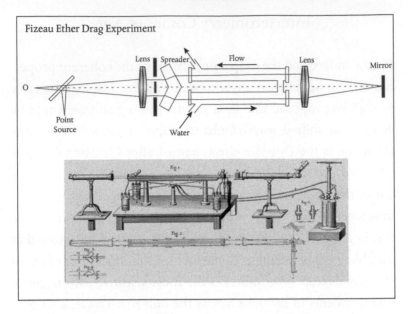

Figure 3.5 Fizeau experiment on light propagating in moving water using an improved Arago interferometer.

Claude Servais Pouillet, *Eléments de physique expérimentale et de météorologie*, sixth edition (Paris: Hachette, 1853), Plate 36 B. See also, pg. 103, É. É. N. Mascart, *Traité d'optique*. Paris: Gauthier-Villars 1889. See also, pg. 52, J. Frercks, "Fizeau's research program on ether drag: A long quest for a publishable experiment," *Physics in Perspective*, vol. 7, (2005): pp. 35–65.

oppositely directed longitudinal flows. The pencil of light traveling down one tube is moving with the flow and is moving against the flow in the other tube. After exiting the glass flanges, the beams are focused to the mirror that redirects the beams back to the lens that recollimates them and directs them down the opposite tube. The two beams are combined at the observation point (labeled "O") where interference fringes are observed. One pencil of light travels with the flow in both arms of the interferometer, while the other pencil of light travels against the flow in both arms, doubling the effect at "O."

When the flow rate was zero, Fizeau saw the fringes in one arrangement, but as the flow increased from zero, the fringes shifted laterally in the view port, and the amount of lateral shift was directly proportional to the speed of the water. Fizeau then reversed the flow, doubling the measured effect. When water was flowing at 7 m per second, he observed nearly a half-fringe shift between the case for flow in one direction versus the other.[22] This half-fringe shift was consistent with a theory that Fresnel had constructed based on the partial entrainment of ether by moving matter, deriving a quantity known as the Fresnel drag coefficient. He had constructed this theory to explain why Arago's measurements of the motion of the Earth using starlight observed through glass prisms had yielded null results. The partial drag of ether by the glass prisms precisely canceled the effect first proposed by Michell. Therefore, Fizeau's results ruled out other hypotheses of complete ether drag as well as completely immobile ether. Fizeau, recognizing the importance of his results, published immediately.[23] Fifty years later, when Einstein was formulating his new view of relativity, he was inspired by the positive experimental results of Fizeau from 1851 more than by the null results that Michelson and Morley would obtain in 1887. In Einstein's hands, Fizeau's experimental results provided support for the relativistic addition of speeds.

When Fizeau published the results of his ether-drag experiment, French experimental optical physics was entering a new and extremely productive phase of research on the physics of light. Fizeau and Foucault were among the last of a genre of serious amateurs who were mostly self-educated outside the university system. At the time when they were working in the 1840s and 1850s in Paris, a similarly self-taught scientist, Michael

Faraday, was at the peak of his career at the Royal Institution. However, in both France and England, young university-educated academics began moving into the top teaching and research positions. They were professional scientists, and their status often eclipsed their older autodidactic colleagues. For instance, when Foucault was passed over for election to the *Société philomathique* in 1849—the event that so disturbed him that he dropped out of public view for half a year—the person who took his slot was Jules Jamin.

Jules Jamin (1818–1886) was a student prodigy at the University of Reims, winning many academic awards and going on to take his first position at the College Bourbon in Caen in Brittany, where fundamental experimental studies of light reflection from metals gained him his election to the *Société philomathique* over Foucault. Jamin was married in 1851 and had two children, his daughter later becoming the wife of Henri Becquerel. In 1852 Jamin obtained a coveted position as a professor at the Ecole Polytechnique and taught the course in general physics, publishing his lecture notes in 1886. It was an influential physics textbook[24] that helped educate the new generation of French physicists at the end of the nineteenth century who would help usher in twentieth-century modern physics.

Jamin continued his experimental studies of light and optics at the Polytechnique. He was inspired by the work of Arago and Fizeau on the interferometric measurements of the refractive index of gases and began to view these measurements as examples of a larger class of measurements that today we call refractometry. He was aware that the point-source and double-slit masks that Fizeau and Arago had used were inefficient, making the interference fringes very dim. Although Fizeau and Foucault used the Fresnel mirror for their long-path-difference interference experiment to solve this problem, the double set of mirrors were difficult to align. In 1856, Jamin conceived of an ingenious new refractometer that had high brightness and was easy to align. The new design dispensed with the double-slit configuration and its relatives that had dominated almost all prior approaches to refractometry.

Jamin knew that light transmitted through a glass plate produced a weak reflection of approximately 4 percent from the first surface. This partial reflection was usually considered a parasitic effect that degraded image quality in optical systems by producing ghost images and glare. Jamin flipped this undesirable background problem to become a new way of performing refractometry, creating what is today called an amplitude-splitting interferometer.

The Jamin interferometer, shown in Figure. 3.6, consists of two thick but extremely flat plates of glass that are silvered on one side. A pencil of light is incident at 45 degrees on a first plate of glass and is partially reflected (path A) and partially transmitted (path B). The transmitted wave is completely reflected by the silvering and exits the glass plate parallel to the first pencil of light but shifted laterally so that two pencils of light are formed that co-propagate toward the second glass plate. The first pencil of light (path A) is transmitted into the second glass plate and completely reflected by the silver. As it exits the glass plate, it is perfectly superposed on the partial reflection of the second pencil of light from the first plate. An exit mask ensures that only the superposition of path A and path B exits the apparatus. The design is ingenious because at the exit mask of the interferometer both beams have traveled exactly the same optical path length

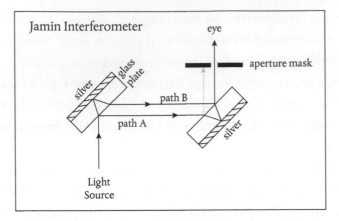

Figure 3.6 The Jamin interferometer uses two identical silvered plates of glass.

(assuming the plates are identical) and have equal intensity. This is the perfect condition to produce high-contrast interference fringes to the eye (or photographic plate). To perform as a refractometer, one simply placed a material in one of the two paths to observe the shift in the fringes.

The importance of Jamin's work was recognized immediately, and two years after publishing it, Jamin received the Rumford Medal from the Royal Society of Britain in 1858. The prestigious award had been endowed by Benjamin Thomson (Count Rumford) years earlier for "an outstandingly important recent discovery in the field of thermal or optical properties of matter made by a scientist working in Europe." Jamin's amplitude-splitting interferometer introduced a new class of refractometer that would culminate, over the next half century, in a vast menagerie of wildly different forms of interferometer, including the most iconic interferometer of them all, invented by a child of the California gold rush—the Michelson interferometer.

4

After the Gold Rush

The Trials of Albert Michelson

... light-waves are now the most convenient means we have for making accurate measurements ... it is the very minuteness of the waves which permits the extraordinary degree of accuracy ...[1]

Albert Michelson (1890)

The birth of modern American science might be dated roughly to the spring of 1881, but the location of the auspicious event was ironically far outside America, situated in the basement of the Potsdam Astrophysical Observatory on the outskirts of Berlin, Germany. This was where Albert Michelson began his first of a series of experiments to measure the motion of the Earth through the luminiferous ether by using an interferometer. It was his interferometer, rather than his attempted measurement of Earth's motion, that would gain for him America's first Nobel Prize and win a new respect by European scientists who, before Michelson, had not taken American science seriously since the days of Benjamin Franklin.

Interference. David D. Nolte, Oxford University Press. © David D. Nolte (2023).
DOI: 10.1093/oso/9780192869760.003.0004

The Prodigy of Calaveras County

The firestorm of 1859 swept through the ramshackle mining town, known as Murphy's Camp, taking one building after another, filling the sky with intense reds and oranges as embers shot skyward like angry stars. Murphy's Camp in Calaveras County in the foothills of the Sierra Nevada mountains in California was named for the Murphy brothers who had crossed the Sierra Nevada mountains with wagons several years prior to the discovery of gold at Sutter's Mill. When the gold rush began, they established claims on Angels Creek and became millionaires selling supplies to local miners. Ten years later the placer mines were still yielding ounces to the pan when the fire impressed a severe form of urban renewal by clearing away the old wood buildings to be replaced with more permanent structures of brick and stone.

In the street, a seven-year-old Albert Abraham Michelson (1852–1931) watched his father's dry goods store go up in flames. Samuel and Rosalie Michelson had settled their family in Murphy's Camp in 1855 after immigrating from Poland by way of the Isthmus of Panama to join Rosalie's sister's family in the town. Rather than mining, Samuel followed the example of the Murphy brothers by selling provisions to the miners, and although he did not become rich, he made enough to support a growing family. To Albert and his sister Pauline, who were born in Poland before they emigrated, were added more siblings who grew up in the rough and tumble world of the mining towns in the foothills of the Sierras where disputes were settled more often with knives and guns than in the courts, and where shop keepers shuttered their shops at night as protection against drunks and thieves.

By the time the precocious Albert was 12 years old, he had outgrown the meager schooling such a town could provide, and his parents shipped him off to San Francisco where he lived with his aunt to attend a boys school that later became Lowell High School, a selective magnet school that is today the oldest public high school west of the Mississippi. He returned home to Murphy's Camp to visit in the summer of 1867, but the placer deposits had finally run out, and the town had gone into decline. Without miners to buy his wares, Samuel could not stay in business, so the family packed up and moved to the new center of mining wealth in the region that had shifted from the gold of the Sierras to the silver of Virginia City, Nevada.

The Comstock Lode of silver ore had been discovered on the side of Mount Davidson in 1859 as the richest lode of precious metal ever discovered. The wealth of the Comstock was so great that it built the financial district of San Francisco and shifted the monetary policies of the United States and in the world. Samuel set up his new dry goods store in Virginia City, and Rosalie kept a close watch over the family growing up in the wild town. Her youngest son Charlie was swept up in the Western life, later living by the gun as a hunter to supply food for mining outfits in Arizona at the height of the guerilla war with Geronimo.[2] But his erudite oldest brother Albert continued his schooling in San Francisco, graduating in 1869 and returning to Virginia City to decide what to do with his life. He had excelled at school and had developed a particular interest in the science of optics, but there were no reasonable prospects for him to continue in that direction.

His father, looking through the local newspaper, the *Territorial Enterprise*, on April 10, 1869, noticed a letter from the Nevada state representative in Congress that a boy from Nevada would be appointed that year to the Annapolis Naval Academy. The selection would be based on an examination open to all boys from the ages of 14 to 18. With encouragement from his father and the headmaster of his high school, Michelson took the exam and tied with two others for first place. The decision went to one of the other boys who was the son of a disabled Civil War veteran. In a strange connection of pop culture to science, the story of Michelson and the Annapolis exam were dramatized in a 1962 episode of the popular TV show *Bonanza* where a family of ranchers, the Cartwrights, living outside Virginia City, helped young Michelson achieve his dreams.

The Jewish community in Virginia City, of which the Michelson family was part, had taken special pride in Michelson's talents and petitioned the congressman to secure for him an extraordinary presidential appointment, of which there were only 10 slots per year. Albert set off alone, bound for Washington on the recently completed Transcontinental Railroad, passing over the location of the "golden spike" that had been driven into a railroad tie of California laurel just a month earlier. He secured an audience with the recently elected president Ulysses S. Grant who regretfully told Michelson that all 10 slots had already been filled for the year. Reluctant to return home in failure so soon after such a long trip, Michelson went directly to Annapolis where he waited three days before the staff relented and finally admitted him for an interview. He was allowed to take the entrance exam, and although he passed handily, there still was no slot for him at the Academy.

Disappointed, he returned to Washington, bought a ticket, and boarded the train for home. Just before the train pulled out of the station, Michelson was shocked to hear someone walking down the aisle of the train car calling out his name. When he announced himself, he discovered it was a messenger from the White House, and he was taken once again to see Grant.[3] It turned out that one of Michelson's examiners at Annapolis was a vice admiral who had convinced Grant to make an exceptional eleventh appointment for this obviously talented student. Michelson received the appointment on June 28 and officially entered Annapolis on June 30, having no time to return home. Within days, he was sailing with the other first-year cadets on the summer training cruise aboard the sloop *Dale* that returned to Annapolis in time for the fall semester to begin on October 1.

Michelson's four years at the Annapolis Naval Academy were unremarkable except for his particular abilities in optics. He made an unintentional enemy of one of his physics teachers by repeatedly failing to read the textbook yet solving optics problems in his own way that the teacher could not easily follow. Given his talents, he was fortunate to be at Annapolis which was one of the very few places in the United States at that time that offered a technical education with such strong emphasis on mathematics, science, and engineering. In 1869, most of the Morrill Act land grant universities that had been set up across the country—such as MIT, Cornell, Purdue, and Berkeley, among others—were still being organized and just beginning to offer classes. Although the Naval Academy was outdated in many aspects, Michelson received one of the best technical educations available at that time in the US. Michelson rose through the student ranks, taking summer training cruises until his fourth year when he began two years at sea. Upon his return in 1875, he was promoted to the rank of Ensign and was appointed to teach physics and chemistry at the Naval Academy.

His appointment as an instructor of physics at Annapolis changed his life and launched him on a trajectory that culminated in the Nobel Prize of 1907. Yet the beginning of this spectacular trajectory was surprisingly prosaic. The teaching of physics had been stuck in rote book learning, but it had recently become more common to perform physics demonstrations in the lecture room following the example set by the Royal Institution in London. As part of the plan to add lecture demonstrations to his classes, Michelson selected a demonstration of the measurement of the speed of light, following Foucault's spinning-mirror technique that he used in 1850 to beat Fizeau for priority measuring the speed of light through stationary water. Foucault used the spinning mirror again in 1862 to measure the speed of light through air to much higher accuracy than Fizeau had achieved in 1849, once again bettering his one-time friend.

In Foucault's apparatus the accuracy of the measurement was limited by how far away the retroflecting mirror could be placed while still returning enough light to be able to see it. The farther the retroflecting mirror was, the dimmer the returned light. This had limited Foucault to 20 m in his 1862 measurement. Michelson had a genius for optical design, and as he studied Foucault's method, he realized that the light-collecting ability of the apparatus could be improved considerably by increasing the focal length and size of the lens and placing it at the center of the apparatus. At the same time, the ease of alignment could be improved by replacing the curved mirror with a flat mirror, shown in Figure. 4.1. With this approach, much larger distances could yield much greater accuracy for the measurement. Therefore, in the process of trying to replicate Foucault's measurement as a simple lecture demonstration, Michelson had hit on a way to surpass Foucault and to measure the speed of light with much higher accuracy.

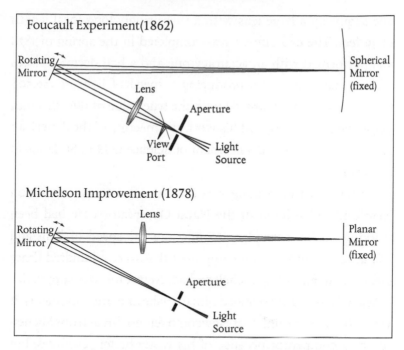

Figure 4.1 Michelson modified Foucault's spinning-mirror system by shifting the lens from a peripheral position to a central position, allowing much longer paths and much greater accuracy for measuring the speed of light.

As Michelson worked to implement his idea, one of his key traits emerged that was to play both a positive and a negative role in his later experimental efforts as well as in his life: he had an obsessive character that made him neglect all else while he was consumed with perfecting a measurement. This has obvious advantages for a scientist making the highest-accuracy measurements of miniscule effects, although it would also come with a severe cost to both health and family. Michelson became absorbed in constructing his apparatus, designing a rotating mirror system that he had custom-built using his own meager money, and acquiring a large lens with a surprisingly long focal length of 39 feet. The experiment was completed in the spring of 1878 and performed with an accuracy one and a half decimal places greater than Foucault's, producing a speed of light of 186,508 miles per second (compared with the true value of 186,282 miles per second). He presented his work at a meeting of the American Association for the Advancement of Science held in St. Louis in August 1878.

Presiding at the meeting was the famous astronomer Simon Newcomb (1835–1909) of the Naval Observatory. He had been working on his own plans to measure the speed of light and had access to money and equipment that were a hundred times greater than anything Michelson had. Newcomb was supportive of Michelson and encouraged him to make finer measurements. It would be easy to think that Newcomb felt no threat from Michelson as a competitor because of his vastly better resources, but his good will toward Michelson was genuine. When Michelson returned from St. Louis, he approached his rich father-in-law for a sizable loan to extend the size and accuracy of his measurement by another order of magnitude. After touring his son-in-law's lab at Annapolis, he was impressed and agreed, and Michelson started construction on a measurement apparatus that spanned nearly half the length of the Annapolis campus. The new system used a lens with a focal length of 150 feet and yielded the most accurate value for the speed of light at 186,355±31 miles per second.[4] The news of this accomplishment was reported by the *New York Times*, bringing Michelson into the limelight as one of America's rising stars in science. His hometown paper in Virginia City picked up the news about the son of a local shopkeeper who was becoming America's most famous scientist.

Michelson's Interferometer

In the late 1870s there were no universities in the United States that awarded advanced degrees in science, so Michelson took a leave of absence from the Navy to study under renowned physicist Hermann von Helmholtz (1821–1894) at the University in Berlin, Germany. In the fall of 1880 he traveled to Germany with his wife and two children and began taking classes, which he generally found unchallenging, especially the tutorial labs in optics because his measurements of the speed of light had honed his skills, making him one of the world's leaders in optical experimental design and practice.

Michelson began to think about an experiment to detect the motion of the Earth through the ether. Maxwell had pointed out only a year earlier, shortly before his death, that measurements of the speed of light around a closed path would show deviations that varied as the squared ratio $(v/c)^2$, where v was the speed of the Earth through the ether. Although Maxwell considered such a ratio to be too small to measure,[5] Michelson disagreed. His experiments on light had convinced him that second-order effects, though small, could yield to experiment, and he began formulating a new method to make the measurement.

The story of how Michelson invented his interferometer is historically obscure and has been the source of debate.[6] It did not seem to happen in a single "Aha" moment but was a series of shifts of thinking on how best to separate beams of light from a single source. He was intimately familiar with Fizeau's system for the measurement of ether drag that used a masked lens to create separate beams, and he knew of Jamin's 1856 method of beam separation in the refractometer that used thick plates of quartz. However, none of these systems were suited to form the widely separated beams that Michelson needed to measure ether drift.

In retrospect, we understand that the key element in the invention of the Michelson interferometer is the partially silvered mirror that acts as a beam splitter. The method to silver mirrors had been perfected by Foucault when he needed to replace the mercury amalgam that ablated from his rapidly spinning mirror, and in 1857 Foucault constructed the first telescopes that used silvered glass mirrors to replace the previous metal mirrors used in Newtonian telescopes. By the late 1870s partially silvered mirrors that could both transmit and reflect light were finding increasing uses, although it seems that no-one before Michelson realized that such partially silvered mirrors could be used to create an interferometer. For instance, Foucault's method for measuring the speed of light used the incidental partial reflection of a tilted glass plate to separate the returning beam from the outgoing beam so that the displacement could be observed in a telescope. In the Foucault arrangement, the partially reflecting glass plate was a peripheral element used for convenience rather than a fundamental and functional part of the design. Michelson had even used this method at Annapolis before his displacements became large enough to allow him to observe the return beam directly without the partial reflections. The key step in Michelson's invention of his interferometer was shifting this *peripheral* element from the viewing port to a *central* location where it separates the beams *and* recombines them, shown in Figure. 4.2. This way of thinking had a similar character to the step he took improving the Foucault speed-of-light system when he moved the lens from a *peripheral* location to a *central* location in the optical design, shown in Figure. 4.1.

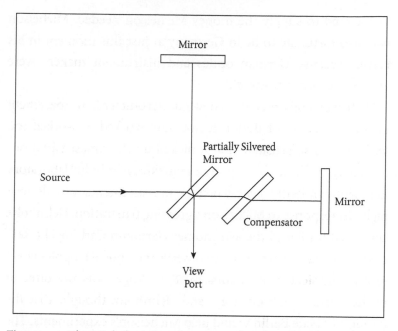

Figure 4.2 The Michelson interferometer placed a partially silvered beam splitter in the center of the apparatus. The compensator added an additional optical path length through glass to match the optical path length of the other beam.

Michelson had already completed his design when he approached Helmholtz in 1880 with a proposal to build the apparatus and test it. Helmholtz was skeptical that it would work, or if it did, he doubted that it would be sensitive enough to detect second-order effects in ether drift. Nonetheless, he was supportive—if Michelson could find the funds necessary to build the delicate optical elements that the experiment would require. Michelson's ties to Newcomb were crucial at this moment as Newcomb brought Michelson to the attention of Alexander Graham Bell who had established a private foundation to provide funds for struggling inventors. Bell was intrigued by Michelson's proposal and agreed to supply the money Michelson needed. Michelson was also fortunate to be in Germany at just this moment in his career, because German optics and instrument makers were among the best in the world.

Michelson constructed his new interferometer in the basement of the University of Berlin. It not only worked, it worked too well—it was so sensitive that it picked up the tiniest vibrations caused by traffic on the city streets outside Helmholtz' laboratory in downtown Berlin, even when the measurements were done at night. In response to Michelson's growing frustration, Helmholtz contacted his friend the astronomer Hermann Carl Vogel (1841–1907), who had been the first to observe the optical Doppler effect in astrophysical observations in 1871.[7] Vogel was the director of the Potsdam Observatory, and Helmholtz thought that the location outside Berlin would help Michelson's experiments. The experiment was transferred to the basement of the Astrophysical Observatory in Potsdam in early 1881.

In the quiet environment in the suburbs, the vibrations were just weak enough to allow Michelson to complete his measurements. He observed the fringes of his experiment as he rotated the apparatus so that the arms of the interferometer changed their orientation relative to the motion of the Earth. As a firm believer in the existence of the ether, Michelson expected to see a shift of one-tenth of a fringe to an accuracy of one-hundredth of a fringe, or what we today would call a signal-to-noise ratio of 10:1. Yet after six months of hard work, during which he took very little rest, he had to conclude that the result of the experiment was null. This conclusion was bittersweet. On the one hand, he had constructed a new metrology device that was the most sensitive apparatus of its kind ever devised, with exquisite sensitivity to the smallest displacements. On the other hand, he had failed to observe the motion of the Earth through the ether.

When he visited important laboratories later during his stay in Europe, Michelson was quick to tell of the performance and prospects of his new invention that he called an *interferential refractometer*, but he was mostly quiet on the results of his ether drift experiments. He was particularly excited to visit Paris because of the strong French tradition in optics that had developed from the work of Fizeau and Foucault. The scientists included not only Jamin but also Marie Alfred Cornu (1841–1902) after whom the Cornu spiral is named, Gabriel Lippmann (1845–1921) who would later receive a Nobel Prize for multilayer dielectrics and was a student of Kirchhoff, and Éleuthere Mascart (1837–1908) whose grandson was Leon Brillouin. Michelson's work was already well known to the French scientists, and Cornu extended a friendly challenge to have him set up a demonstration of his interferometer in the lab at the Ecole Polytechnique. Cornu was almost gleeful as Michelson initially failed to get fringes to emerge from a makeshift interferometer that he constructed using a candle as its light source. Michelson did not know at the time that a candle is the worst possible choice as a coherent light source. After several days without success, Cornu offered to let Michelson withdraw with the honor of having tried his best, but Michelson would not have it. Consistent with his obsessive character he forged ahead and finally got the fringes to appear when he achieved perfect zero-path difference between the two arms of the interferometer. When Cornu finally was shown the fringes he exclaimed, "My friend, you have it," and he called the physics faculty to see the effect for themselves.[8]

Michelson stayed in Paris for several months working and discussing optical physics with his hosts. Although he was enjoying himself, Michelson was becoming anxious to settle his future. He had been in Europe for nearly two years, dragging his wife and children after him as he moved from Berlin to Heidelberg to Paris. Michelson's scientific reputation within the United States had been growing, helped along by influential supporters like the editor of the *American Journal of Science* who had published Michelson's first journal paper. Finally, in the spring of 1882 he received an offer to join the faculty of the newly founded Case School of Applied Science in Cleveland, Ohio. The Case Board of Trustees offered Michelson a position as Instructor of Physics at $2,000 per year with a start-up package of $7,500 to equip the new laboratory.[9] Relieved and excited, Michelson resigned his Naval appointment in April and sailed for New York to prepare for his new faculty position that started September 1, 1882. He set up his first laboratory in the family barn on the Case property.

Personal Firestorms

During the late colonial period of US history, a wide swath of northeastern Ohio was owned by the little colony of Connecticut for no other reason than that it lay at the same latitude despite the interposing large colony of New York and the proprietorship of Pennsylvania. This western wilderness was "reserved" by royal charter for the people of Connecticut should the day come when Connecticut became too crowded to hold all its citizens. The area was called the "Western Reserve." After the Revolutionary War, the Northwest Ordinance of 1787 opened up the Northwest

Territories to settlement, and Connecticut ceded the region to a land speculation group led by General Moses Cleveland who surveyed the area in 1796 and founded a city at the mouth of the Cuyahoga River. By 1826, settlement in the Western Reserve (many of the settlers actually came from Connecticut) had progressed to the point that a college of higher learning was needed, and the Western Reserve College and Preparatory School was founded in the small town of Hudson, Ohio about 30 miles south of the new city of Cleveland. It was called the Yale of the West and became known for its progressiveness, especially in the cause of abolition.

In 1882 the philanthropist Amasa Stone donated a large area of parkland on the eastern side of Cleveland to attract Western Reserve College to relocate to Cleveland. Stone's grant also provided adjacent land for Case College. A few years later, Main Hall of Case College was completed next to Western Reserve in the area that came to be called University Circle. It is located at the southern extreme of a green parkway that meanders alongside Doan Brook that flows north into Lake Erie.[10] This was a big improvement over the old Case family home in downtown Cleveland, and Michelson was happy to move his sensitive interferometry laboratory from the barn into the basement of Main Hall in 1885.

That same year, Michelson traveled to Montreal to attend a meeting of the British Association for the Advancement of Science, presided over by William Thomson (Lord Kelvin) and by Lord Rayleigh, where he fortuitously met Edward Morley of Western Reserve College who also was attending the meeting. They were pleased to find that they were soon to be neighbors and struck up a friendly acquaintance. After Montreal, Thomson gave a series of lectures at Johns Hopkins University in Baltimore on the current state of physics, and both Michelson and Morley attended. At the end of the lectures they took the same train back to Cleveland and had deep discussions the whole way. Morley was the elder experienced scientist, strong in math and chemistry, while Michelson was the junior faculty adept at experimental physics and eager for the benefit of Morley's experience. At the conference, Rayleigh encouraged Michelson to continue his experiments on the detection of the motion through the ether, sensing that these were among the most crucial experiments in physics, and Michelson decided it was time to begin again despite his previous null results. Morley volunteered to assist.

The central conundrum of the physics of the ether was the coexistence of two conflicting experimental conclusions on ether drag. The first was the Fizeau experiment on moving water that supported a stationary ether with *partial drag*, while the other was the Michelson experiment of 1881 in Potsdam that supported George Stokes' theory of a fluid ether with *complete drag*. The two conclusions could not both be correct, and Michelson was confident of his 1881 results, so he decided to repeat the Fizeau experiment of 1851 on the speed of light in moving water. He designed a new experimental configuration that would be much more sensitive than Fizeau's approach. Michelson and Morley began their work together in Michelson's laboratory in mid-1885.

It is fortunate that Morley had the patience of a saint, because Michelson was difficult to work with—demanding and at times caustic. It didn't help that Michelson became obsessed with the experiment, unable to sleep at night as he turned the experimental details over and over in his mind. His physical and mental conditions began to deteriorate under the self-imposed stress, and his family life unraveled as he became unfeeling toward his wife Margaret and harsh to his children. Michelson's mental state collapsed under severe exhaustion. At that time, such a nervous breakdown was viewed as a form of insanity caused by actual physical damage to the brain—called a softening of the brain—from which one did not recover. Margaret, at a breaking point herself after years living with the obsessive behavior of her husband, committed him, against his will, to the care of a New York doctor with the full expectation that he would be put away for good into an asylum. Michelson was forced to leave his faculty position, though he did not resign. Case, in turn, did not fire him but promptly replaced him with a new lecturer in physics.

In a strange twist of fate, the doctor to whom Margaret entrusted his care was a rare visionary ahead of his time. Dr. Allen Hamilton, a grandson of Alexander Hamilton, was a nerve specialist who believed that many mental disorders had behavioral origins rather than physical. Hamilton's work, predating Sigmund Freud by about a decade, helped to change the treatment guidelines for mental patients across the United States. He immediately recognized that Michelson's problem was behavioral—actually exhaustion and obsessiveness—and he promised him a rapid recovery, news that came as a somewhat unwelcome surprise to Margaret who had already begun to feel a sense of freedom while separated from her husband. Michelson recovered quickly under Hamilton's treatment, but when he returned to Case in the fall of 1885, he had to accept half pay to make up for the salary that had been promised by Case to his replacement.

Although he resumed living in the family household, he remained emotionally absent to his wife who he felt had betrayed him, and he moved into a separate room of the house. Margaret would divorce him many years later, citing mental cruelty, and during the divorce proceedings, his children were carefully coached to paint him as a monster. As his family left the courtroom, all he could say to them was that they had not heard his side of the story. Margaret left the next day with full custody of the children—Michelson never saw some of them again. This sad story stands in stark contrast to the memories recounted by his daughter Dorothy Michelson from his second marriage who appears to have had a close and loving relationship with her father.[11]

Now back at the benchtop—the best remedy for any true experimentalist—work on the Fizeau experiment resumed with the help of Morley who had stood by Michelson through the ordeal. Morley had augmented the equipment during Michelson's absence in preparation for the final series of measurements. These were performed in late 1885 and early 1886, using the experimental configuration shown in Figure. 4.3, to a much greater degree of detail and precision than Fizeau had reported in 1851—confirming the Fresnel drag coefficient. Their results showed that the speed of light was not just the sum of the speed of light in a dielectric plus the speed of the dielectric motion (Galilean relativity), but instead was a fractional increase just as predicted by Fresnel. Michelson was especially pleased to obtain a definite and finite result, as opposed to his somewhat disappointing null results of Potsdam.

Michelson and Morley published their results in the May 1886 issue of the *American Journal of Science*.[12] Although confirming Fizeau's previous results did not cause a stir, it further established Michelson as the world leader in precision measurements of the speed of light. It also highlighted the usefulness of his interferometer for metrology purposes. Someone who took notice

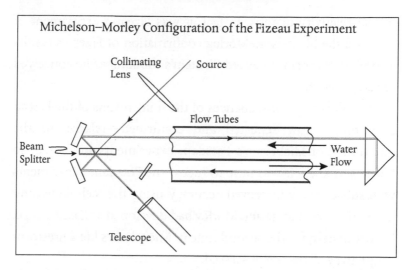

Figure 4.3 The Fizeau flow experiment using Michelson's modified interferometric configuration. (If Michelson had rotated the table during the experiment, he would have discovered the Sagnac effect.)

was Woldemar Voigt of the University of Göttingen in Germany who realized that the equations for light propagating through the moving water in the Fizeau and the latest Michelson–Morley experiment had an invariant structure if they were rescaled by a factor given by $\sqrt{1 - \beta^2}$, where β is the ratio of the speed of light in the dielectric to the speed of light in vacuum.[13] Furthermore, by performing a coordinate transformation of the wave equation using this factor, he showed that the time coordinate became mixed with the spatial coordinate. Voigt came dangerously close to preempting Einstein's later theory of special relativity, publishing his work in early 1887[14] a full two years before Fitzgerald proposed his relativistic length contraction, five years before Lorentz proposed his first theory of the electrodynamics of the electron, and 18 years before Einstein published his theory of special relativity of 1905. In fact, it was Fizeau's experiment of 1851, and the Michelson–Morley confirmation of Fizeau's results in 1886, that most influenced Einstein's thinking as he conceived of his special theory.

The conflicting consequences of the finite results of the Fizeau experiment—now repeated and confirmed—relative to the null results of Michelson's Potsdam experiments still stood, so Michelson knew it was time to repeat the Potsdam experiments with substantially improved accuracy using the techniques and the equipment that he and Morley had developed at Case College. Unfortunately, for the second time in Michelson's life a firestorm swept away all that he possessed.

At 2:30 a.m. on October 27, 1886 a fire broke out in the tower of Main Hall, followed by a deafening explosion that shook the students out of their beds in the nearby Adelbert College dormitory of Western Reserve. The flames quickly engulfed the upper stories of the building, fanned by a strong wind that drove sparks against a wooden barn across the street that also caught fire. When the joists under the slate roof burned through, the entire roof crashed down into the building with a deafening roar, destroying everything inside including all the books of the founder's library as well as Michelson's equipment in the basement. By morning, only smoke and steam escaped from the blackened skeleton of the main building of Case College.[15]

The Michelson–Morley Experiment

Fortunately, all was not lost. At the time of the catastrophe, Michelson and Morley had been working together for more than a year, and Morley's lab at Western Reserve still held some of their equipment, so they decided to rebuild in his lab. The new attempt on the ether-drift experiment had two key advances over the Potsdam experiment. The first was the new interferometer design by Michelson that crisscrossed the light beams back and forth across the stone table, shown in Figure. 4.4, creating an interferometer that performed as though it was ten times larger than the table itself, with a sensitivity ten times better than at Potsdam. The other key element of such a sensitive experiment was an ingenious method proposed by Morley that provided isolation from even the tiniest vibrations—the entire experiment could float in a bath of mercury. For this, they found a solid floor in the basement of the Adelbert dormitory of Western Reserve. They constructed a cast-iron tub to hold the mercury, and the 5 ft-square stone table supporting the optical elements was placed on a donut-shaped wood pallet that floated on the mercury. The 2-ton apparatus could be set in a slow spin with a single push by a finger, rotating without friction or disturbance.

Half a year passed as they assembled the necessary funds to buy the equipment and to set up the large apparatus. On July 8, 9, 11, and 12, 1887 they performed the final measurements at noon and at 6 p.m. each day as the rotation of the surface of the Earth added to the motion of the Earth around the sun by different amounts. One of them would set the table spinning as the other walked slowly with the moving table, watching unblinking at the fringes through the one-quarter-inch aperture of the telescope. If their attention on the fringes faltered for even a moment, they had to restart that trial. The goal was to observe the smallest possible shift of the interference fringes as the table spun through a 90-degree angle. If the ether were stationary relative to

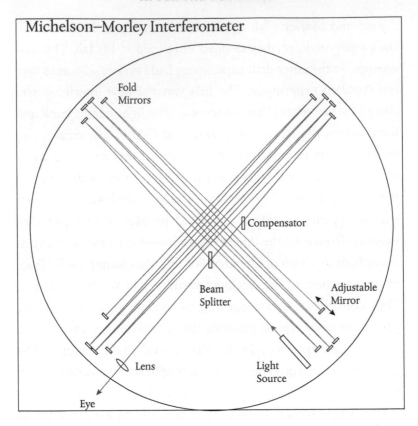

Michelson–Morley Interferometer

Fold
Mirrors

Compensator

Beam
Splitter

Adjustable
Mirror

Lens

Light
Source

Eye

Figure 4.4 The multiple-folded-path layout of the Michelson–Morley experiment created an interferometer that was effectively ten times larger than its footprint.

the Earth—as the Fizeau experiment implied—then the fringes should have shifted by almost a half fringe during the rotation, representing a 50-percent shift. However, what Michelson and Morley observed after many tries was a deviation of the fringes by no more than 2 percent with a mean value of only 1 percent. In the end the results were incontrovertible. To the greatest experimental precision yet achieved, using the most advanced metrology in the world at that time, they concluded with confidence that there was no detectable motion of the Earth through the luminiferous ether.

Michelson was visibly upset with the results. He had proceeded with the conviction that at least some vestige of motion through the ether would emerge from such an exquisitely sensitive measurement. The apparatus he and Morley had constructed was so sensitive that the hoof-falls of horses walking down the street outside the building could be detected as slight blurring of the fringes. Coupled to this level of sensitivity, the speeds of the Earth's rotation and revolution about the sun are extremely high and are even measurable fractions of the speed of light. It was almost unconceivable that such large motions could not be detected to the smallest degree. But those were the facts. The speed of light measured in the laboratory was independent of any absolute motion of the Earth through the heavens.

Everyone who knew Michelson, and knew about his recent history with mental exhaustion, feared for his well-being. Margaret urged him to go away with the family on a vacation to distance himself from his disappointment. He obliged, only to find on their return home that their house staff had stolen all of Margaret's jewelry and heirlooms and had run off. He left Margaret to work with the police to track down the culprits as he and Morley wrote up their results for publication. The paper appeared in the November 1887 issue of the *American Journal of Science*.[16] This time the paper did cause a stir. Lord Rayleigh and Hendrik Lorentz had fully expected the experiment to succeed. However, there was no question about the validity of Michelson's and Morley's experimental results—the detail and care as described in the paper were unprecedented, and the results were accepted beyond any doubt.

This gave Rayleigh and Lorentz and everyone else working on the physics of the ether no wiggle room. Motion through the ether was truly undetectable, and any theory of the dielectric properties of matter had to deal with that fact.

The subsequent history and consequences of the Michelson–Morley experiment are well known. Eugene Fitzgerald explained away the results by proposing that the length of the arms of the interferometer contracted by just the right amount—no more than a few tens of nanometers—in the direction of motion to cancel the effects of the Earth's passage through the ether. Lorentz independently arrived at the same conclusion, not as an ad hoc postulate, but as a consequence of his theory of the electron. Coordinate transformations emerged from Lorentz' work that were remarkably similar to the conclusions Voigt had made before Michelson and Morley had even begun their ether drift experiments. Poincaré, in turn, recognized these coordinate transformations as abstract rotations in four-dimensional coordinate space that mixed space and time, although he did not take the final step to conceive of space-time as Minkowski did in 1908. It was Einstein with his relativity postulates of 1905 that set the matter to rest—there was no ether and no need of the ether. The speed of light was a fundamental property of nature and invariant to any relative motion. This fresh perspective gave both a trivial and a profound explanation for the null results of the Michelson–Morley experiment.

Yet Michelson's most valuable legacy was the invention of the interferometer and the incredible sensitivity it possesses. The detection of gravitational waves at the Laser Interferometric Gravitational Observatory (LIGO) in 2015, arguably the smallest physical effect ever measured by mankind, was accomplished using a Michelson interferometer (Chapter 6). Furthermore, Michelson himself would turn his eyes to the heavens and use his interferometer to measure the size of stars. But before following Michelson into the field of astronomy, we will take a short look at the descendants of Michelson's interferometer that emerged during the first golden age of interferometry in the decades following Michelson's invention.

A Menagerie of Interferometers

There are broadly two classes of interferometers. These are amplitude-splitting interferometers and wavefront-splitting interferometers. In an amplitude-splitting interferometer, a wave is split into a transmitted and a reflected wave, such as by partial reflection and transmission at a dielectric interface, and the waves are later combined to interfere. As an example, interference within thin films is a form of amplitude-splitting interferometer. In contrast, in a wavefront-splitting interferometer, the parts of waves that interfere come from different points on the same wavefront, traversing different paths to arrive at the detection plane where they interfere. Young's double-slit experiment is an example of a wavefront-splitting interferometer.

Despite such a "clean" classification system of the two types, there are so many different ways that waves can take different paths to eventually recombine and interfere, that a wild menagerie of interferometer designs has sprung up over the past century. Many of these hybrids combine aspects of both amplitude-splitting and wavefront-splitting forms, making them difficult to classify. Because there are so many different types of interferometer, they tend to be given the names of their inventors as a way of keeping them straight, but this practice obscures similarities. For instance, there are several "named" interferometers of the Michelson type that have only slight differences based on their different uses. The nuanced differences include such modifications as lenses or purposeful misalignments such as tilting of elements to generate parallel fringes. Thus, one has the Twyman–Green interferometer that is a Michelson-type interferometer used for curved optical component testing to be contrasted with the Kösters interferometer that is a Twyman–Green for measuring optical flats. Similarly, there are numerous Jamin-type interferometers that have been adapted for interference microscopy, such as the Sirks–Pringsheim interferometer that is an off-axis Jamin interferometer that can be used for interference microscopy. In turn, there are many forms of interference and phase-contrast microscopes that represent an entire sub-field to itself.[17]

One interferometer design deserves special mention because of the key role it plays in many modern applications. This is the Fabry–Perot interferometer, named after Charles Fabry (1867–1945) and Alfred Perot (1863–1925), who were professors of physics at the University of Marseilles. In the early 1890s they began working together on the measurement of small gaps between metal surfaces. Fabry, the younger member of the team and more theoretically inclined, had done his PhD thesis at the University of Paris on the study of interference fringes and brought this expertise to the problem. Perot was a master of laboratory techniques and had already distinguished himself in the measurement of thermodynamic properties of vapor phases. With their complementary talents, Fabry and Perot worked fruitfully together for over a decade before administrative positions pulled them in different directions.

Their key contribution to the study of interferometry was the introduction of a pair of partially silvered parallel plates that faced each other. Michelson had already introduced one such plate into his interferometer of 1880 in Berlin that acted as both the beam splitter and the beam combiner. In the early 1890s Ludwig Mach (the son of Ernst Mach) and, independently, Ludwig Zehnder separated these two functions of splitting and combining by using two partially silvered plates in what came to be called the Mach–Zehnder interferometer. The interference fringes in the Mach–Zehnder interferometer have the same two-wave interference properties as Michelson fringes, but the separation of the beams into separate paths makes it more adaptable to phase manipulation, making it one of the most common interferometer configurations used in modern optics today. However, both the Michelson and the Mach–Zehnder, as well as the Jamin and all its variants, despite their exquisite sensitivity and proven usefulness for metrology, have purely sinusoidal interference fringes, making it difficult to locate fringe maxima and minima precisely. The sinusoidal fringes are the result of two-wave interference from two alternative paths that light takes through the interferometer. However, as Fabry and Perot explored the interference properties of their two opposed silvered plates, they found new interference properties that relied on multiple waves interfering with precisely prescribed phases.

The properties of multi-wave interference had already been explored theoretically by the British astronomer George Bidell Airy (1801–1892) in the second edition of his mathematical textbook published in 1831.[18] He considered the multiple reflections of light waves that occur in a plane parallel dielectric slab that produce an infinite number of successively smaller reflections and transmissions that add up to produce a reflected and transmitted intensity that varies as a function of the slab thickness and also of the angle of view. The result of the infinite sum was a finite and periodic function that later was called the Airy function. He predicted that the interference would be observed as rings of varying brightness when viewing a distant light source through a glass plate because of the slightly different path lengths taken by rays at different angles. The Austrian geophysicist Wilhelm Haidinger (1795–1871) was the first to observe very faint Airy rings in 1855 when studying the optical properties of thin minerals like sheets of mica. Because the partial reflection at the surfaces of transparent materials is only moderate (glass produces only a 4 percent partial reflection), the interference fringes from thin films and plates have very small contrast compared with the 100 percent contrast that is achievable in a Michelson interferometer. This was the situation when Fabry and Perot began working together: two-wave Michelson and Mach–Zehnder interference had the advantage of maximum contrast but weak fringe localization, while multi-wave interference seemed to have little advantage because it had both weak contrast and weak fringe localization—until Fabry and Perot changed the rules of the game.

Airy's derivation of his function assumes reflection and transmission coefficients at surfaces of the slab, r and t, that are not tied to the underlying refractive index of the slab material itself. If the reflection and transmission are purely due to Fresnel coefficients, then the fringes from thin films would have been doomed to small contrast and little usefulness. But by using partially silvered surfaces, Fabry and Perot could achieve surface reflection coefficients up to 90 percent. Under such strong reflections the Airy function

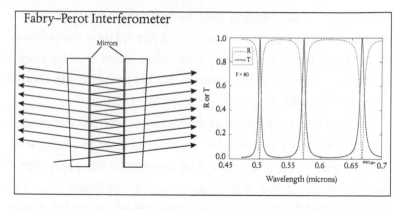

Figure 4.5 Fabry–Perot Etalon composed of two partially silvered plates facing each other. The interferometer is illuminated by a broad source, but only a single ray is shown to illustrate the multiple reflections and transmissions. The transmission for a high-finesse cavity can approach unity with very narrow fringes in monochromatic light.

becomes strongly non-sinusoidal, as shown in Figure. 4.5, with sharp and pronounced transmission peaks separated by deep dark regions, creating both high contrast and fine fringe localization. They called this unusual property of their new interferometer its "finesse."

Fabry and Perot realized the unique value of this new type of interferometer, not only for their original metrology problem, but also for spectroscopy. At that time, the highest resolution spectrometers used finely ruled diffraction gratings perfected by the American physicist Henry Rowland at Johns Hopkins University (Chapter 7). However, the high finesse of the Fabry–Perot interferometer provided higher resolution than possible with ruled gratings, and over the next decade Fabry and Perot explored laboratory and astrophysical applications of their interferometer, including ultra-precise measurements of Doppler shifts in gases and moving objects. With the new emphasis on atomic spectroscopy that was driving the quantum revolution of the early twentieth century, the Fabry–Perot interferometer became a mainstay in laboratories. It also provided the central resonator for lasers later in the century (see Chapter 8).

The timeline of key events in the history of interferometers and interferometry is given in Table 4.1, and the main interferometer designs are shown in Fig. 4.6. The strength of the Fabry–Perot interferometer is the ability to measure extremely fine distances in the laboratory. However, the versatility of Michelson's interferometer positioned it to revolutionize measurements of astronomical distances in the wider universe, as discussed in the next chapter.

Table 4.1: Historical timeline of key events in interferometry and interferometers.

1665   Robert Hooke
        In Robert Hook's *Micrographia* (1665) he described colors in thin
        plates and soap films and guessed that they might have something
        to do with vibrations related to the thickness of the films.

1665   Francesco Maria Grimaldi
        Writing in *Physico mathesis de lumine, coloribus, et iride, aliisque annexis
        libri duo* (1665) Grimaldi described how light scattered from a small
        aperture produced fringes, coining the term "diffraction."

1704   Isaac Newton
        Newton's influential *Opticks* (1704) was the result of several decades
        of experiments and theory (dating back to 1665) including his obser-
        vation of circular colored rings around the point of contact of a
        spherical glass surface with a flat plate. He related his transmis-
        sion "fits" to the thickness of the air film, but his particles of light
        did not possess an intrinsic length scale that could explain the
        phenomenon.

1785   David Rittenhouse
        Rittenhouse placed 50 hairs between two finely threaded screws
        with an approximate spacing of about 100 lines per inch and
        observed multiple bright lines from this first demonstration of a
        diffraction grating. He described the effect in *An optical problem, pro-
        posed by Mr. Hopkinson, and solved by Mr. Rittenhouse* in the *Transactions
        of the American Philosophical Society*.

1804   Thomas Young
        As he prepared physical demonstrations for his lectures on physics
        at the Royal Institution, Young observed the diffraction pattern
        from a thin obstruction in a light beam. He illustrated interference
        from a double slit in a beautiful plate in his *Course of Lectures on Natu-
        ral Philosophy* (1807). This is arguably the first interferometer though
        he did not use it as a measurement apparatus (Chapter 1).

1816   Francois Arago and August Fresnel
        During the period from 1816 to 1818 Arago and Fresnel worked
        together in the laboratory of the Paris Observatory. The double-
        beam interferometer they developed to measure the refractive index
        of moist air was the first true "interferential refractometer" (Chap-
        ter 2). This same design was used thirty years later by Fizeau and
        Foucault (Chapter 3).

1816   Augustin Fresnel
        In addition to deriving the first diffraction integral, Fresnel
        devised two related approaches to divide the wavefront and form
        interference patterns: the double mirror and the biprism.

1819   Siméon Poisson
        Although he remained an "emissionist," Poisson was one of the
        first to embrace Fresnel's mathematics, predicting the "Spot of
        Arago" (Chapter 2) as well as explaining the dark fringes arising
        from multiple reflections inside a thin glass plate.

1827   John Herschel
        In his *Treatise on Light*, Herschel described dim fringes that appear
        when light from an extended source reflects from a thin glass plate.
        The fringes arise from the interferences among multiple internal
        reflections.

Continued

Table 4.1: *Continued*

1831   George Biddell Airy
The Astronomer Royal derived the results of an infinite series of
partially reflected waves inside a glass slab and obtained what is
today called the Airy function. In 1835 he derived the Airy rings in
the diffraction pattern from a circular aperture.

1834   Humphrey Lloyd
Lloyd proposed an alternative to Fresnel's double mirrors.

1845   Armand Hippolyte Fizeau and Léon Foucault
Interference of light with long path differences was made using
Fresnel's mirrors and a dispersive prism in the first demonstration
of spectral interferometry (Chapter 3).

1850   Léon Foucault
Foucault gains priority over Fizeau in the measurement of the speed
of light in air relative to water using an Arago interferometer. The
slower speed of light in water confirms the wave nature of light
against the emissionists' theory of light.

1851   Hippolyte Fizeau
Fizeau measures the speed of light in moving water using an Arago
interferometer and confirms Fresnel's drag coefficient (Chapter 3).
This hallmark experiment established one of the key results that
ultimately led Einstein to the special theory of relativity.

1855   Wilhelm Karl Haidinger
Haidinger observes circular interference rings in thin mica sheets,
now known as Haidinger fringes, confirming Airy's theory of
multi-wave interference in thin slabs.

1856   Jules Celestin Jamin
Jamin invents the double-plate refractometer with high fringe visibility and ease of alignment to measure the refractive index of
materials and gases (Chapter 4).

1858   Léon Foucault
Foucault devised a knife-edge technique for optical component
testing. This is a form of phase-contrast imaging that visualizes
phase errors in the surface of a silvered mirror.

1862   Hippolyte Fizeau
Fizeau developed an interferometer for optical component testing that relied on uniform fringes (Fizeau fringes) to designate ideal manufacturing. (It appears that Fizeau and Foucault could not stop competing against each other in the field of optical interferometry.)

1881   Albert Michelson
Michelson invented the interferential refractometer using a partially silvered beam splitter and performed the first ether-drift experiment in the basement of the Potsdam Observatory (Chapter 4). Gravitational waves were detected in 2015 using the largest Michelson interferometers ever constructed with optical arm lengths of 4 km.

1890   Albert Michelson
Michelson proposes the astronomical interferometer and measures the diameters of Jupiter's moons the following year (Chapter 5). His design is strikingly similar to a proposal made by Fizeau in 1868 and attempted by Stephan in 1873.

1891   Ludwig Mach and Ludwig Zehnder
The Mach–Zehnder interferometer was proposed first by Zehnder and refined by Mach (the son of Ernst Mach). The Mach–Zehnder (MZ) interferometer combines the best features of the Jamin and Michelson interferometers. The MZ interferometer has eclipsed the Michelson interferometer to become the most common interferometer used in modern non-linear optics, atomic and molecular optics, and quantum optics systems.

1896   Lord Rayleigh
Rayleigh used an Arago interferometer to measure the refractive index of gases. This interferometer configuration is sometimes called a Rayleigh interferometer although it was developed 80 years earlier by Arago and Fresnel.

1899   Charles Fabry and Alfred Perot
Fabry and Perot perfected multiple-wave interferometry (dating back to Hooke and colored fringes in thin films) by separating the reflecting surfaces of a glass plate into separate high-reflectance coatings known as etalons (Chapter 4). The Fabry–Perot interferometer performs as a high-resolution spectrometer or as an optical resonator for lasers.

Continued

Table 4.1: *Continued*

1913	Georges Sagnac In the Sagnac interferometer, two beams counter-propagate around a loop. If the loop is rotating, a phase shift is acquired between the two beams that is proportional to the angular velocity. When constructed using fiber optics, this interferometer can be used as an optical gyroscope for navigation.
1932	Frits Zernike The phase contrast microscope was invented by Zernike as a means to image phase objects (Chapter 7). He won the 1953 Nobel Prize in Physics for his invention.
1948	Denis Gabor Holographic recording in photographic emulsions and wavefront reconstruction were demonstrated by Gabor as part of a two-step process to improve microscope imaging resolution (Chapter 8). Gabor was awarded the Nobel Prize in 1971.
1949	André Mirau Mirau invented a compact interferometric microscope objective that incorporates an in-line Michelson-type interferometer for phase-contrast imaging.
1952	Georges Nomarski Another form of phase-contrast microscopy uses a birefringent prism to create two shifted lateral optical paths that interrogate a semi-transparent sample and are recombined for viewing. This is the archetype of image-shearing phase-contrast microscopes.
1956	Robert Hanbury-Brown and Richard Twiss Intensity-based interferometry introduced a radically new paradigm and was later used for stellar interferometry (Chapter 5). The first demonstrations using classical waves have developed into a variety of modern quantum optical interferometers (Chapter 9).

A Managerie of Interferometers

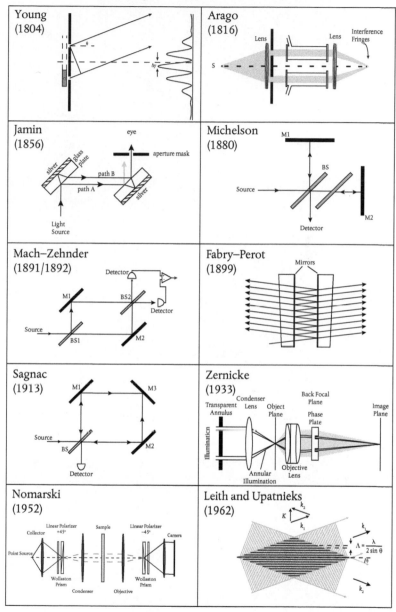

5

Stellar Interference

Measuring the Stars

... the greatest need of stellar astronomy at the present day ...
is some means of measuring the apparent angular diameters of
stars[1]

Sir Arthur Eddington (1920)

Astrometry is the precision side of *astronomy*. It answers the
kinematic questions of astrophysics like: Where? How far?
How big? How fast? For instance, the search by James Bradley
(1693–1762) for stellar parallax in 1728 was a problem in astrometry. He found instead stellar aberration, which is the effect of
the motion of the Earth through space on the apparent location
of stars—also an astrometry problem. The first stellar parallax
was measured by Friedrich Bessel (1784–1846), of Bessel function
fame, for the binary star system known as 61 Cygni in 1838, with a
parallax of about 300 mas.[2] The parsec, an astronomical distance
scale, is the distance to an astronomic object that produces 1
arcsecond of parallax, placing 61 Cygni a bit over 3 parsecs away
from Earth. One parsec is a distance of 3.26 light-years.

Once the astrometric problem of stellar parallax had been
solved, at least for nearby stars, the next serious question in

Interference. David D. Nolte, Oxford University Press. © David D. Nolte (2023).
DOI: 10.1093/oso/9780192869760.003.0005

astrometry was how big are they? By the early twentieth century, our sun was the only star whose size was known, although many theoretical estimates on star sizes were floating around. Arthur Eddington, the leading theorist of stellar evolution at that time, placed the problem of the size of stars as "the greatest need of stellar astronomy." The obstacle to measuring the size of stars lies in the sizes of the optical elements of telescope. In 1920, the largest telescope, the Hooker on Mt. Wilson in California with an impressive diameter of 2.5 m, was not big enough to resolve the disk of any star in the sky. Ironically, the answer to the problem of measuring something as big as a star was *not* making something big enough, but rather finding something small enough to measure it with. The answer came in the smallness of the wavelength of light—a millionth of a meter—that allowed vast astronomical sizes to be brought down to Earth.

The Moons of Jupiter

Armand Hippolyte Louis Fizeau was perhaps the first physicist who truly grasped the power of optical interferometry and who understood how to design general interferometric metrology systems. Although Fizeau measured the speed of light without using interferometry,[3] he became an expert in interference effects, and he was one of the first to recognize that each emitting region of a light source was coherent with itself and thus could produce interference. Fizeau used his understanding of coherence to improve on Arago's and Fresnel's original refractometer, and he became adept at thinking of alternative interferometric designs and their uses. In an address to the French Physical Society in 1868 he suggested that a double-slit mask could be used on a telescope to determine sizes of distant astronomical objects.[4] The optical design of this stellar interferometer was similar to his refractometer, but instead of a local light source, a distant star provided the illumination, passing through the two slits in the objective lens of a telescope where the observation of the fringes would be the same as in the refractometer. Yet despite the similarity in optical layout, Fizeau's crucial insight was that a distant bright object, like a star, would behave like his local source to provide the coherence needed to see fringes in the image. This idea is remarkable, tapping into the deep physics of spatial coherence, concepts that challenge students and practitioners even today. Fizeau knew from his work with his refractometer that the size of the light source relative to the separation of the slits affected the visibility of the fringes. Therefore, by direct analogy, the size of the star relative to the separation of the slits would affect the fringe contrast in the same way.

Édouard Stephan (1837–1923), at the observatory outside Marseilles, made his first attempt to use Fizeau's configuration in astronomical observations in 1873 by observing the star Sirius. When the slits were added to the objective lens, he was excited to see interference fringes across the point-spread-function of the star, but the slits could not be brought far enough apart to see any change in the fringe contrast (which is needed to estimate the size of the stellar disk). Sirius was too small and too far away for the interference contrast to change as the slits were increasingly separated across the 65 cm diameter of the objective lens of his telescope. At that time, the distance to Sirius was only roughly known through parallax to be around 3 parsecs away,[5] and its size was completely unknown, so there was no way for Stephan to know whether the diameter of his telescope was sufficient to see the effect.[6] Stephan could only set an upper limit to the size of Sirius, and the results were buried in a short note to the Comptes Rendus.[7]

Seven years later, Michelson's visit to Paris in 1881 set up one of the prime questions in the history of interferometry—was Michelson told of Fizeau's stellar interferometer? It is unlikely that Fizeau would have been present. After Fizeau's wife died early in their marriage, he spent most of his time in seclusion at his home outside Paris. Michelson was well aware of Fizeau's published research and acknowledged him as a direct inspiration of his own work in interference effects. Hence, if he had met him, he would have mentioned it, as he mentioned so many others like Lippmann, Cornu, and Mascart in his letters. Therefore, it can be concluded with some confidence that Michelson and Fizeau never met. On the other hand, the question of whether Michelson may have been told of Fizeau's stellar interferometer, maybe just in passing, is a mystery that remains lost to the shrouds of history unless or until additional evidence comes to light.

Michelson's discovery (or rediscovery) of stellar interferometry owed a bit of its success to the "failed" Michel–Morley experiment of 1887. At the conclusion of that experiment, both he and Morley were disappointed by their null results—so much so that they scrapped their plans to make additional measurements three months and six months later when the motion of the Earth would be oriented at 90 degrees and then 180 degrees to its starting point during its orbit around the sun. Yet the interferometer had performed with such extreme precision that Michelson began thinking of measurements that went beyond the luminiferous ether. This is when his mind turned to the idea of using interferometry to measure the size of astronomical objects. His understanding of interference effects and the role of coherence now went beyond Fizeau's original 1868 conjecture, and Michelson had the mathematical skill to derive the equations to prove the idea.

Michelson's rising fame within the American physics community gave him mobility, and he moved to the newly founded Clark University in Worcester, Massachusetts in 1889. Once Michelson was established at Clark, he began his first experimental studies of spatial coherence to mimic stellar interferometry. He drilled pairs of small holes in aluminum sheets, illuminated either by a sodium flame or by a white light source, using different spacings and placing them different distances from the objective lens of a 4-inch-diameter telescope at the input to his interferometer. He also mathematically derived the effect of the holes on fringe visibility and published this preliminary work in 1890 in a paper titled "On the application of interference methods to astronomical measurements."[8] The paper presented the first mathematical description of the role of spatial coherence on interference fringe visibility, and he proposed to use it in a stellar interferometer. The general principle of stellar interferometry is illustrated in Figure. 5.1.

Once Michelson was convinced about what he was looking for, he approached the Harvard College Observatory in Cambridge with a proposal to try his technique on the 15-inch-diameter refractor housed there. He began his observations in October 1890, but several weeks of bad weather and unstable atmosphere prevented him from reliably measuring the fringe visibility from any star. Rather than being discouraged at yet another null result, he recognized the problem of the turbulent atmosphere and

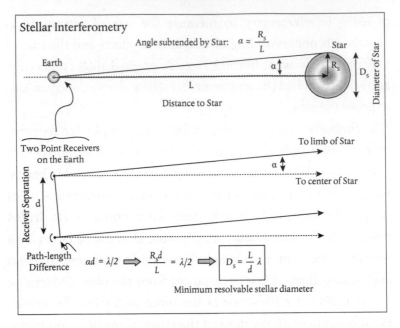

Figure 5.1 Principle of stellar interferometry. The star with a radius R_S is a distance L from Earth and subtends an angle α. Two point receivers separated by a distance d on the Earth detect a wave originating from the limb of the star tilted at the angle α that produces the path difference αd as in the Young double-slit configuration. When this path difference is approximately a half wavelength, destructive interference cancels the common signal. In first-order interferometry this quenches the fringe visibility. In second-order (Hanbury Brown–Twiss) interferometry this cancels the noise correlations.

turned to an observatory that was famous for its still skies—the Lick Observatory on Mt. Hamilton overlooking San Jose, California. An added benefit of spending time at the Lick Observatory was a chance to return to the beloved state of his youth.

Michelson designed and built a double-slit appendage to fit the 12-inch telescope at Lick and traveled to California in July 1891. Although his ultimate goal was to measure the diameters of stars,

Michelson decided to begin with objects closer to home—the Galilean moons of Jupiter. There were several reasons to try his technique on the moons. First, the sizes of the moons were already approximately known. Although the moons could not be fully resolved by the ground-based telescopes of the day, deviations from the point spread function of a point source allowed estimates of their diameters, and the values obtained by interferometry could be compared against them. Second, the superior accuracy of interferometry would provide better values for the sizes of the moons, which was a topic of keen interest for many astronomers. Michelson made observations on four nights in August using the modified 12-inch telescope while simultaneous observations were made using a 16-inch telescope, also located at Kitt Peak, to measure the point spread functions of the direct images. The interferometric diameters matched well with the deconvolved direct images, and Michelson published the results later that year,[9] gaining media attention as he proposed to make the first ever measurements of stellar diameters.

At just this moment, Michelson's attention was pulled away from stellar interferometry and turned to a platinum-iridium meter bar housed at the *Bureau international des poids et mesures* (BIPM) in Paris, France. He had realized that the precision of his interferometer could relate the size of the meter bar, the physical basis of the metric system, to something much more fundamental, such as the wavelength of the red cadmium spectral line. In the fall of 1891, he traveled to Paris with a special-purpose interferometer and over the ensuing year made increasingly accurate measurements of the meter. The meter bar itself required tremendous care and handling in a temperature-controlled room to prevent any thermal drift in length and any corrosion of the bar. Michelson's achievement was to convert this fragile physical meter bar to an exact number of wavelengths that was a fundamental and immutable property of nature. The final value for the length of the meter was 1.553 million wavelengths of cadmium red.

When Michelson returned from Paris in 1893, he had been poached once again, moving from Clark University to the newly founded University of Chicago, where he was the first head of the department of physics. It was at Chicago that he began interacting with some of the top minds of the day. One of the first assistant professors he hired into the department was Robert A. Millikan (1868–1953), and one of his colleagues at Chicago was the astronomer George Ellery Hale (1868–1938) who had great plans for advancing astronomy in the United States. Although he had hoped to return to stellar interferometry, success on the meter-bar measurements and other events in his research life diverted his attention for almost 30 years before he once again took up the challenge of measuring stellar diameters. In the meantime, others noticed Michelson's 1891 publication on astronomical interferometry. One of these was a peripatetic astronomy doctoral student in Germany.

Karl Schwarzschild (1873–1916) was born in Frankfurt, Germany, shortly after the Franco-Prussian War heralded the rise of Prussia as a major political force in Europe. By the time Schwarzschild was 16, he had taught himself the mathematics of celestial mechanics to such a degree that he published two papers on the orbits of binary stars. He also became fascinated in astronomy and purchased lenses and other materials to construct his own telescope. After a short time at the University of Strasbourg (part of the German Federation after the war) Schwarzschild transferred to the University of Munich where he studied under Hugo von Seeliger (1849–1924), the premier German astronomer of the day.

When von Seeliger and Schwarzschild became aware of Michelson's theoretical analysis of stellar interferometry in 1890 and of his use of an interferometer to measure the size of Jupiter's moons in 1891, they realized that replacing the double-slit mask with a multiple-slit mask would make the widths of the interference maxima much narrower and would cause a star to produce a multiple set of images on the image plane of the telescope associated with the multiple diffraction orders. More interestingly, if the target were a binary star, the diffraction would produce two sets of diffraction maxima—a double image! If the "finesse" of the grating were high enough, the binary star separation could be resolved as a doublet in the diffraction pattern at the image, and the separation could be measured, giving the angular separation of the two stars of the binary system.

Schwarzschild enlisted the help of a fine German instrument maker to create a multiple-slit system that had an adjustable slit separation. The device is shown in Figure. 5.2 from Schwarzschild's 1896 publication on the use of the stellar interferometer to measure the separation of binary stars.[10] The device is ingenious. By rotating the chain around the gear on the right-hand side of the apparatus, the two metal plates (with four slits each) could be raised or lowered, causing the projection onto the objective plane to have variable slit spacings. In the operation of the telescope, the changing height of the slits does not matter, because they are near a conjugate optical plane (the entrance pupil) of

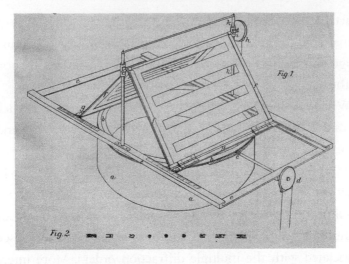

Figure 5.2 Illustration from Schwarzschild's 1896 paper describing an improvement of the Michelson interferometer for measuring the separation of binary star systems.

the optical system. Using this adjustable multiple-slit system, Schwarzschild and two colleagues made multiple observations of well-known binary star systems, and they calculated the star separations.

Schwarzschild's publication demonstrated the first systematic use of stellar interferometry. This major achievement was performed before he had received his doctoral degree—and on a topic orthogonal to his dissertation topic. Yet this fact is virtually unknown to the broader physics community outside of astronomy. If he had not become so famous later for his solution of Einstein's field equations, Schwarzschild nonetheless might have been famous for his early contributions to stellar interferometry.[11] Despite the success of the astronomical interferometer for measuring the separation of binary stars, the goal of using the method to measure the diameters of single stars remained elusive.

Betelgeuse and the Interferometrist

The star called Betelgeuse is living too fast and too hard, and its life will be short. As a red giant, it will grow so obese that its girth will no longer be supported by its inner fuel, and it will implode in a spectacular supernova that will shine almost as bright as the moon in a night sky about 100,000 years from now. Then it will shrink from sight, collapsing into a neutron star.

Until the Hubble Space Telescope was launched in 1990 no star had ever been resolved as a direct image. Within a year of its launch, using its spectacular resolving power, the Hubble optics resolved—just barely[12]—the red supergiant Betelgeuse. No other star (other than the sun and one star in the Southern Hemisphere[13]) is close enough or big enough to allow images of its stellar disk, even for the Hubble far above our atmosphere. The reason is that the diameter of the optical lenses and mirrors of the Hubble—as big as they are at 2.4 m diameter—still produce a diffraction pattern that smears the star's image. Yet, as Fizeau first pointed out, and Michelson confirmed, information on the size of a distant object is encoded as phase in the light waves that are emitted from the object, and this phase information is accessible to interferometry.

In November 1907 Michelson was notified by the Swedish Academy that he had been selected to receive that year's Nobel Prize in physics. Unlike today when it is announced to the world at the same time that the awardee is notified, in Michelson's day the awardee was strictly forbidden from telling anyone, even as they were required to arrange travel to Sweden to receive the prize. At the same time, Michelson was awarded the prestigious Copley Prize of the Royal Society, so he traveled to England with plans to extend his trip to Oslo afterward. When his travel plans became known at the award event of the Royal Society, he was hard pressed to fend off questions by well-wishers who gave him winks and knowing nods.

Michelson's excitement about traveling to Oslo to attend the famous Nobel gala was dampened when word came that King Oscar II (whose birthday in 1890 had been celebrated by a mathematics competition that launched Henri Poincaré to fame) had died. As the city went into mourning, all of the festivities surrounding the awards were canceled. Michelson's low mood as he traveled by passenger liner to Copenhagen from London was lowered further when he encountered Rudyard Kipling on board who was that year's Nobel Prize awardee in literature. Initially pleased to meet the famous author, Kipling proceeded to bash America and Americans, and Michelson, a proud ex-Navy man, promptly turned his back on him.

Nonetheless, the Swedes were congenial, and Michelson's time in Oslo was enjoyable. At the end of his Nobel public lecture, as he was answering questions from audience members who gathered around him, a young man approached who said in a lowered voice, "You don't know me. I am your son."[14] Michelson was shocked to find himself face to face with his oldest son, his namesake Albert, whom he had not seen since the sad day when his family had left him at the divorce court 11 years before. Albert had been in Italy when he learned that his father was receiving the Nobel Prize, and he had traveled to Oslo. This meeting began a rapprochement between Michelson and his two sons by his first marriage, although his daughter from that marriage never reconciled.

Michelson returned to Chicago in triumph, lending prestige to both the university and his physics department. His own research efforts expanded over the next 10 years, revolving around optical metrology that went beyond interferometry to include high-precision spectroscopy. As Michelson was approaching retirement, and was beginning to envision a life of more leisure, the United States entered World War I. He was reactivated into the Naval Reserve and called to Washington to help improve Naval range finders. In Washington, he reconnected with Hale, who had left Chicago in 1904 to build the observatory on Mt. Wilson in California in association with the newly founded California Institute of Technology, later to be known as Caltech, which he hoped to fashion into a West-Coast version of the University of Chicago. Hale had also been called to Washington, like Michelson, and the two met frequently for lunch at the exclusive Cosmos Club on Massachusetts Avenue. With Michelson as a captive audience, Hale suggested that he should split his time in retirement between the Mt. Wilson Observatory and Caltech.

George Hale could be highly persuasive. He was a natural salesman, charismatic and personable, full of energy, with a gift for opening the eyes of people to see what they hadn't known they wanted. Hale used this talent to convince philanthropic organizations to build successively larger astronomical observatories, starting with Yerkes outside Chicago, then Mt. Wilson outside Los Angeles, and finally his crowning achievement, Palomar outside San Diego. He did this at a time when the US government was not in the business of investing in science, so he approached Carnegie and others like him to help build his dreams. He also was a community servant, organizing the National Research Council (NRC) of the National Academies, which enlists academics from across the country to help specify the state-of-the-art in science and to identify gaps in technology that require national investment to support national missions.

Hale must have been an engaging lunch companion. He had wide-ranging personal interests that included Egyptian archeology—he was present when Howard Carter opened the tomb of King Tutankhamen in the Valley of Kings in 1922. While in Washington, Hale was also having lunch with Robert Millikan, Michelson's younger colleague at Chicago, who he convinced to become the head of the physics department at Caltech. Hale struck a chord with Michelson who had always intended to return to stellar interferometry. In the 30 years that he had been away from the subject, no one else had yet succeeded in measuring the diameter of a star. Hale offered him the use of the Mt. Wilson staff and facilities, and Michelson agreed to visit Pasadena after the war.

When Michelson traveled to California in 1919, first to visit the University of California at Berkeley and then to Caltech, there was no consensus on the size of stars, and hence there was no guarantee that a stellar interferometer would be of any value to astronomy. In 1890, when Michelson first proposed measuring stellar discs, it was generally assumed that all stars were about the same size as the sun, although there was no specific evidence to support this. However, in 1906 the Danish chemist and astronomer Ejnar Hertzsprung (1873–1967) used Planck's blackbody measurements, in conjunction with spectral measurements made at the Harvard College Observatory by Antonia Maury, to conclude that some of the brighter red stars in the sky could be immense—tens or hundreds of times larger than the sun. During the same time, Henry Noris Russell (1877–1957), from Princeton University, used eclipsing binaries to estimate the density of stars and discovered a wide range of densities. Some red stars had exceedingly low density, which suggested very large size. Both Hertzsprung and Russell began plotting the brightness of stars as a function of their color. Rather than being a random splay of points, they found that the stars clumped into "bands" on the graphs that suggested relationships between brightness and color. These types of plots later became known as Hertzsprung–Russell diagrams and are today a primary tool for understanding stellar evolution to get a rough idea of stellar size.

During Michelson's first visit to Pasadena in August 1919, he was introduced to the staff at the Mt. Wilson Observatory. The list reads like a Who's Who of American astronomy: John Anderson, James Ritchey, Frederick Seares, Francis Pease, Adriaan van Maanen, Edwin Hubble, Milton Humason, and Harlow Shapley. Later that year Anderson used Michelson interferometry to measure the separation of spectroscopic binaries, building on the earlier work by Karl Schwarzschild. Based on the success of Anderson's measurements, Hale notified Michelson, then back at Chicago, at the end of the year that the stellar diameter project would proceed using the 100-inch Hooker Cassegrain reflecting telescope.

The Hooker Telescope was a marvel. Promoted by Hale and funded by John Hooker and Andrew Carnegie, the primary mirror for the Hooker weighed 4.5 tons and took six years to polish. It had been carefully transported on a Mack truck, inching up the unpaved winding mountain-road access to the observatory site on top of Mt. Wilson. When it was installed in 1917, it was the largest, and hence the most powerful, telescope in the world (until it was eclipsed by the Hale telescope constructed at Palomar in 1949). The Hooker was exceedingly delicate, and Hale would risk only a 20-foot metal beam at its peak, because the weight of the appendage could flex the structure. This 20-foot beam set the scale for the smallest subtended angle the interferometer could measure. Using the simple formula $d = L\lambda/D$ and plugging in $d = 20$ feet gives an angle $D/L \approx 20$ mas as an estimate of a measurable angle for the proposed stellar interferometer. Stars too small or too far away would not be measurable. The big question was whether any star would be close enough and big enough to see the fringe visibility vanish when the mirrors of the interferometer were at their greatest extent allowed by the size of the metal beam.

As work began building the Michelson stellar interferometer in the summer of 1920, there was still no agreement on which stars would be the best candidates to observe. Capella, Sirius, Antares, and Betelgeuse were all considered, but initially the distances needed for the mirror separations were estimated to be over 100 feet, which were much farther than the 20-foot beam that the Hooker could bear, so Michelson began calculations on partial extinction of fringe visibility to consider how much of a change in visibility could be reliably measured. During his traditional summer stay, Michelson worked with Francis Pease (1881–1938) making initial fringe visibility tests with prototypes of the instrument, and they detected clear fringes from distant stars for the entire span of mirror separations.

Michelson had just returned to Chicago to teach fall classes in 1920 when he received a letter from Hale relating a paper Arthur Eddington had presented at the August 24 meeting of the British Association. Eddington, the world expert on stellar evolution, had recently revised his estimates on stellar sizes. The opening lines of his address stated: "Probably the greatest need of stellar astronomy at the present day, in order to make sure that our theoretical deductions are starting on the right lines, is some means of measuring the apparent angular diameters of stars."[15] He then pointed to the star Betelgeuse as the star with the largest angular magnitude, estimating the angle to be 51 mas.[16] This was just within the capability of the 20-foot steel beam on the Hooker. Hale enthusiastically informed Michelson that Pease and John Anderson would proceed with observations of Betelgeuse.

The configuration of the Michelson stellar interferometer is shown in Figure. 5.3. Light from a distant star is intercepted by the two mirrors M1 and M4 positioned on the steel beam on the top of the telescope tube and are reflected into the collecting optics of the telescope. When the two pencils of light intersect at the view port, they produce interference fringes, just as in the Fizeau refractometer. Because of the finite size of the stellar source and, given the separation between the mirrors M1 and M4, the fringes are partially washed out. By adjusting the mirror locations, a certain separation can be found where the fringes completely wash out. The size of the star is then related to the separation of the slits for which the fringe visibility vanishes.

After several months of preliminary trials, observations were made on the night of December 13, 1920. Because the mirrors on the beam could not be adjusted from the ground, a wood plank was placed across the aperture on top of the telescope where an assistant sat to move the mirrors and adjust an optical wedge at the shouted commands of the observer 50 feet below. It was tedious work, but when the telescope was first aimed at Algol, fringes could be seen across the full span of mirror separations. The mirrors were then set at 121 inches (based on Eddington's estimate of the size of Betelgeuse) and the telescope was aimed at the red giant in Orion's shoulder. The outer fringes vanished! According to Eddington, who later received a letter from Pease

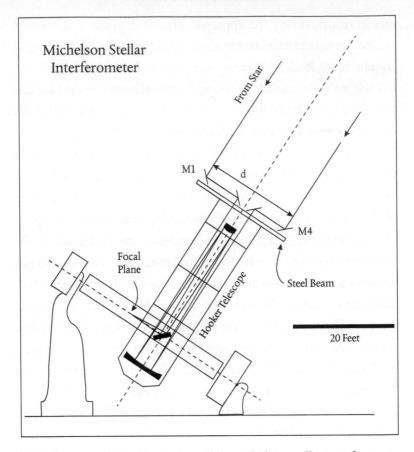

Figure 5.3 Optical configuration of the Michelson stellar interferometer on the Hooker Telescope. Fringes at the view port are partially washed out by the finite size of the star. By adjusting the mirror separation d, the fringes can be made to vanish entirely, yielding an equation that can be solved for the size of the star.

describing the event,[17] around midnight "the observers adjourned to the 'galley' for cocoa and toast with their heads full of sums— for the lack of fringes to Betelgeuse was presumably the real thing." Pease recounted in the letter that, "Presently Anderson speaks up and says, 'It's a whale of a thing—let's see, as big as the orbit of Mars.'" The deduced diameter was 45 mas, which was very close to Eddington's estimate.[18]

The success of the measurements once again put Michelson in the limelight. His house in Chicago was mobbed by reporters, and because he was absent, they interviewed his teenage daughters instead. Once the excitement died down, Michelson moved on to other topics as he usually did, leaving the field of stellar interferometry to Pease. Over the next decade Pease made measurements on virtually all the red giants that were visible from Pasadena,[19] adding stellar sizes to the Hertzsprung–Russell diagram, and where parallax measurements were available, with distances too. Pease built a 50-foot version to extend the range of the interferometer, measuring β Andromeda to 16 mas with a mirror separation of 30 feet. But the weight caused too much flexure in the overloaded Hooker, and only a few additional stars were added to the catalog.

In that moment on December 13, 1920, the range of human probes of the cosmos had jumped to new scales, and a brave new era of astronomy seemed in the offing. But soon all the potential targets had been exhausted and once again human efforts to bring the skies down to Earth had hit a barrier. As with all technological advances, rapid development is usually followed by saturation— until a new technology arrives to begin the rapid development cycle again. This was also the case with stellar interferometry. Michelson's entire career as an interferometrist was based on the interference of wave *amplitudes*, but it is possible to extend the principle to wave *intensities* by which much larger separations of detectors, and much greater distances in the heavens, can be accessed.

The Intensity of Hanbury Brown and Twiss

Only a few years after Einstein proposed the existence of the photon,[20] an undergraduate student at Cambridge University, Geoffrey Taylor, performed and published[21] an iconic quantum experiment: Young's double-slit experiment performed one photon at a time. To achieve such low photon fluxes, Taylor attenuated the light from an arc lamp by passing it through multiple plates of smoked glass, illuminating the double slits, and making long exposures of the interference pattern on photographic plates. The attenuated light intensity was so low that in one of his experiments the exposure lasted three months. Quantum physics was so new at that time that Taylor could only estimate the average number of photons in his light beam, but he succeeded in performing the experiment under conditions where only one photon, on average, was in flight between the source and the photographic plates at any given time.

When Taylor developed his plates, he saw Young's familiar interference fringes. In the face of Einstein's conclusion that a photon was absorbed as an indivisible quantum of light energy, Taylor's experiment showed that the single photon in his apparatus still succeeded in interfering with itself as if it had passed through both slits at the same time. This is a central conundrum of quantum physics—the display of wave-particle duality. The photon is detected as a whole particle, exciting one molecule at a time in the silver halide film, while behaving as a wave as it passes through both slits. We know today that it is the quantum wavefunction of the photon that samples both slits, and the squared amplitude of the wavefunction at the locations on the photographic plate determines the probability for photon absorption.

Because the wavefunction samples all accessible parts of the apparatus, the wavefunctions from the two slits interfere and create absorption fringes on the photographic plate.

Despite this wave-particle duality, or perhaps because of it, the predictions of quantum mechanics for the double slit are identical to the predictions of classical electromagnetic theory. This is because both quantum and classical theories predict interference in the wave amplitudes, regardless of whether they are electric field amplitudes or quantum amplitudes. Therefore, it is not necessary to invoke the photon, or even quantum mechanics, to explain the resulting bright and dark fringes on the photographic plate. Quantum effects don't manifest themselves under these "first-order" experimental conditions that treat only amplitudes. To see quantum effects, which are due to the quantized arrival of photons at the plate, it is necessary to look beyond just averages and look instead at the fluctuating photon arrivals.

The arrival of a stream of photons at some surface is analogous to the pitter-patter of rain on a skylight or on a tent flap during a camping trip. Each raindrop is discrete, producing a soft "ping" or "plop" as it hits. If it's not raining too hard, you can hear the individual plop of each drop, and when there are many drops, the loudness varies, fluctuating, now louder, now softer, as more or fewer drops hit the tent at any given time. These fluctuations in the discrete arrival times are known as "shot noise" and is a fundamental property shared by all discrete phenomena, whether they are classical raindrops or quantum particles like photons. Therefore, to gain access to the quantum behavior of light requires second-order experiments that measure fluctuations. But even in second-order experiments, the manifestations of quantum phenomenon are still subtle, as evidenced by an intense controversy that was launched by experiments performed in the 1950s by a radio astronomer.

Robert Hanbury Brown (1916–2002) was born in Aruvankandu, India, the son of a British army officer. He did not seem destined for great things, receiving an unremarkable education that led to a degree in radio engineering from a technical college in 1935. He hoped to get a PhD in radio technology, and he even received a scholarship to study at Imperial College in London, when he was urged by the rector of the university, Sir Henry Tizard, to give up his plans and report immediately to a mysterious job interview at the Air Ministry. Hanbury Brown did not know it, but the winds of war were rising over Europe, and England had just embarked on a secret mission to track enemy aircraft by using radio waves—what later came to be known as radar. There was an appalling lack of expertise in radio engineering in England, especially in the Air Ministry, and Tizard was acting as a talent scout, poaching students from his own university to steer toward the project.

Tizard was a hard man to ignore, and without knowing what he was getting into, Hanbury Brown took the interview at the Ministry and was inspected by Robert Alexander Watson Watt, the recent inventor of radar, who spent more time talking than interviewing without ever telling Hanbury Brown anything at all about the job. Two weeks after the interview, Hanbury Brown received an offer with a shockingly low salary of 214 pounds sterling per year. Against his better judgment, he decided to take it, and within days he was reporting to an architecturally incongruous fairy-tale castle known as Bawdsey Manor in a remote corner of Suffolk overlooking the English Channel. The team that Watson Watt assembled at Bawdsey was a misfit crew of physicists and technicians who knew very little about radio engineering in the real world.[22] Hanbury Brown fit right in and began the most exciting and unnerving five years of his life. He was smack in the middle of the early development of radar defense, leading up to the crucial role it played in the Battle of Britain in 1940 and the Blitz from 1940 to 1941. Partly due to the success of radar, Hitler halted nighttime raids in the spring of 1941, and England escaped invasion. Hanbury Brown was then sent to the Naval Research Lab in Washington DC where he helped in the American development of radar transponders.

Returning to Britain after the war, the now knighted Sir Watson Watt, of whom it was said he could sell a refrigerator to an Eskimo, talked Hanbury Brown into joining a research consulting firm he was setting up. The work went well and kept him busy until Sir Watson Watt decided to move to Canada to marry his mistress, leaving Hanbury Brown suddenly unemployed. In 1949, 14 years after he had originally planned to start his PhD, he enrolled at the relatively ripe age of 33 at the University of Manchester. Because of his background in radar, his faculty advisor told him to look into the new field of radio astronomy that was just getting started, and Manchester was a major player because it administrated the Jodrell Bank Observatory, which was one of the first and largest radio astronomy observatories in the world. Hanbury Brown was soon applying all he had learned about radar transmitters and receivers to the new field, focusing particularly on aspects of radio interferometry.

Radio interferometry obeys the same physics principles as Michelson's light-based stellar interferometry. After Michelson's collaborator Francis Pease tried, and mostly failed, to get reliable data from the 50-foot interferometer on the Hooker Telescope, visible stellar interferometry had gone dormant in the early 1930s. This was because, despite Michelson's initial hope that interferometry would be insensitive to atmospheric turbulence, the first-order interference on which stellar interferometry relies cannot avoid fluctuations in the phases of the incoming light fields, presenting a fundamental limitation to this kind of interferometry. Radio telescopes, on the other hand, were mainly directional detectors rather than imagers, so some of the limitations of optical interferometry were not so severe for radio interferometry.

Martin Ryle (1918–1984), at Cambridge with Derek Vonberg (1921–2015), developed the first radio interferometer and measured the angular size of the sun[23] and of radio sources on the sun's surface that were related to sunspots.[24] Despite the success of their measurements on the star nearest to the Earth, their small interferometer was unable to measure the size of other astronomical sources. From Michelson's formula for stellar interferometry, longer baselines between two separated receivers are required to measure smaller angular sizes. For his PhD project, Hanbury Brown was given the task of designing a radio interferometer to resolve the two strongest radio sources in the sky, Cygnus A and Cassiopeia A, whose angular sizes were unknown. As he started the project, he was confronted with the problem of distributing a stable reference signal to receivers that might be very far apart, maybe even thousands of kilometers, a problem that had no easy solution.

After grappling with this technical problem for months without success, late one night in 1949 Hanbury Brown had an epiphany.[25] He wondered what if, instead of measuring the radio fields and phases, the two separate radio antennas measured only intensities? Everyone knew that the intensity in a radio telescope fluctuates in time like random noise. If that random noise were measured at two separated receivers while trained on a common source, would those noise patterns look the same? After a few days of considering this question, he convinced himself that the noise would indeed share some common features, and the degree to which the noise was similar should depend on the size of the source and the distance between the two receivers, just like Michelson's fringe visibility. But his arguments were back-of-the-envelope, so he set out to find someone with the mathematical skills to do it more rigorously. He found Richard Twiss.

Richard Quentin Twiss (1920–2005), like Hanbury Brown, was born in India to British parents but had followed a more prestigious educational path, taking the Mathematical Tripos exam at Cambridge in 1941 and receiving his PhD from MIT in the United States in 1949. He had just returned to England, joining the research division of the armed services located north of London, when he received a call from Hanbury Brown at the Jodrell Bank radio astronomy laboratory in Manchester. Twiss traveled to meet Hanbury Brown in Manchester, who put him up in his flat in the neighboring town of Wilmslow. The two set up the mathematical assumptions behind the new "intensity interferometer" and worked late into the night. When Hanbury Brown finally went to bed, Twiss was still figuring the numbers. The next morning, the tall and lanky Twiss appeared in his silk dressing gown in the kitchen and told Hanbury Brown, "This idea of yours is no good, it doesn't work"[26]—it would never be strong enough to detect the intensity from stars. However, after haggling over the details of some of the integrals, Hanbury Brown, and then finally Twiss, became convinced that the effect was real.

Rather than fringe visibility, it was the correlation coefficient between two noise signals that would depend on the joint sizes of the source and receiver in a way that captured the same information as Michelson's first-order fringe visibility. But because no coherent reference wave was needed for interferometric mixing, this new approach could be carried out across very large baseline distances.

Hanbury Brown and Twiss (the duo hereafter abbreviated as HBT) settled the mathematical theory and enlisted the radio telescopes at Jodrell Bank to measure the radio sources in the sun, just as Ryle had done with his radio field-based interferometry. The HBT intensity interferometer worked precisely as predicted, so they launched the next phase of their research program to measure the angular sizes of the Cygnus and Cassiopeia sources. In 1952, using one standing radio telescope at Jodrell Bank and loading a second, smaller, telescope into the back of a lorry truck, they began their first set of observations. The truck was parked in a farmyard about a mile away from Jodrell Bank and all went well. The truck then worked its way across Cheshire, farmyard to farmyard, until the noise correlations had vanished. Once the data were analyzed, HBT had measured for the first time the angular sizes of both the Cygnus and the Cassiopeia sources, publishing the results in a paper titled "A New Type of Interferometer for Use in Radio Astronomy" in 1952.[27]

To understand why intensity interferometry works, it helps to think about what starlight from a single star looks like at the surface of the Earth. Even though the properties of starlight are about as far from laser light as you can imagine, the vast distance to the star allows some of the light to act like laser light. Laser light is coherent, meaning that the phases (locations of the peaks and troughs of the wave) at one part of a light beam are related to the phase at others. The distance to the star allows the light from each part of the star to spread out and behave like a partially coherent laser beam. Because the angular size of the star is small but finite, the waves coming from one part of the star are out of phase with the waves from other parts, creating a random pattern of high and low brightness caused by random interference effects known as a *speckle pattern*, illustrated in Figure. 5.4. If two receivers are side-by-side, they are within the same speckle and see the same noisy signal. But if one of the receivers is moved farther away, its noise signal no longer perfectly tracks with the fixed receiver, and as the distance gets even bigger, the signals look completely different. The distance d at which the two signals are only half

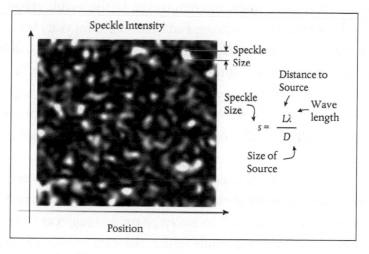

Figure 5.4 Speckle intensity as a function of position, representing the instantaneous intensity cast by starlight on the Earth. The speckle size is related to the distance to the star, the wavelength, and the size of the star. By measuring the size of the speckle, and knowing the distance to the source, the size of the star is obtained.

correlated is related to the size of the source through the familiar expression from Michelson $d \approx L\lambda/D$, where L is the distance to the star, λ is the wavelength of the radio wave, and D is the source diameter. Although the HBT effect depends only on intensity, the expression for the angular resolution is approximately the same as for Michelson's field-based interferometer.

After the experimental success of the new intensity interferometer, HBT confirmed that the noise signals they were comparing were largely immune to the wild fluctuations in the ionosphere, unlike Michelson-type radio interferometers. The fluctuating phases of the arriving signal did not appear (to lowest order) in the detected intensities. This welcome side effect of the radio experiments made HBT turn their attention to Michelson's visible-light stellar interferometer that *was* limited by the turbulent atmosphere. Stellar interferometry in the visible range of the electromagnetic spectrum had been stalled for years because most stars were too small and too far away to resolve, even using interferometry, amid atmospheric turbulence. However, if the new intensity interferometer could work in the visible range of frequencies, then the sensitivity to atmospheric turbulence could be removed, and stellar optical interferometers would gain new life.

To test the principle, Hanbury Brown set up his first optical interferometer in the dark room at Jodrell Bank in late 1955. The key question, going back to his epiphany in 1949, was whether the intensity noise of a visible source, detected by two separated detectors sharing the same path length to the source, would be the same. In the laboratory experiment, the physical size of the source was too large to test the other (and maybe more important) question about measuring the angular size of sources using intensity interferometry, but answering the correlation question was a first step. After the usual difficulties with equipment, the experiment successfully showed strong correlation of the two detector signals when their apertures were superposed on the source (subtending identical angles). Then, as one of the detectors was shifted laterally, shown in Figure. 5.5, the correlations slowly decreased, ultimately vanishing as a function of the shift, following Twiss' theoretical predictions.[28]

The purpose of the first optical HBT experiment was simply to gain confidence before gearing up for a demonstration of intensity

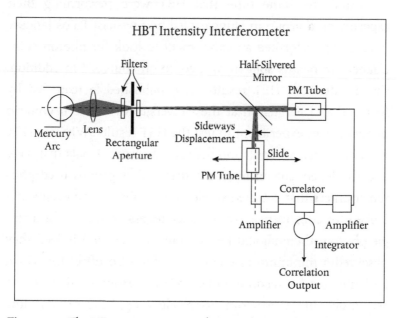

Figure 5.5 The HBT experiment on the correlation of optical intensity fluctuations. The photomultiplier (PM) tubes detect the fluctuating arrival of photons, and the signals are correlated at the output. When the optical path lengths from the source to the receivers are the same, positive correlations in the detected signals are observed.

interferometry to measure the angular size of stars. Instead, the publication of their paper in January 1956 started a fire storm of criticism that persisted for years that centered on the quantum nature of the photon and on whether HBT had indeed measured what they claimed.[29] The argument *against* the correlation that HBT had observed is very simple when only a single photon is present at a time in the apparatus. Because photons are discrete and indivisible, if a photon arrives at one of the detectors, then it cannot arrive at the other separate detector, and vice versa. Therefore, only one detector should fire at a time—exactly the opposite effect of what HBT observed.

Around the same time that HBT were performing their experiment, a group in Hungary led by physicist Lajos Jánossy (1912–1978) performed an experiment to look for photon coincidences in beams split from a common source.[30] In addition, shortly after the HBT results were published, a team led by Brannon and Fergusson at the University of Western Ontario performed an experiment to test the HBT results. Whereas HBT had performed a correlation between detector signals that were linearly dependent on intensity, these other groups used photon counters that detected single photon absorption events and looked for coincidence. By the discreteness argument for a single photon, there should be no coincidences, and indeed they observed a minimum in coincidence for zero offset for which HBT measured a maximum in correlation. How could these two seemingly similar experiments produce opposite results—low coincidence but high correlation?

This question drew in several noted physicists, most of whom argued against the HBT results, questioning the validity of the HBT experiment. If Einstein's photon is indivisible, then they argued that the correlations between the two detectors should follow the coincidences and hence should vanish. Many suspected that HBT had measured something else, probably some systematic in their experiment that they had not considered. Reflecting back on this barrage of criticism, Hanbury Brown wrote: "If science had a Pope we would have been excommunicated."[31] But ideas can become entrenched to the point that prejudice trumps knowing better. The error that the physicists were making was thinking too rigidly about photons as discrete particles. Photons are indistinguishable and must satisfy properties of quantum statistics—statistics that the physicists knew well—but their knowledge was overshadowed by their mental picture of discrete photons arriving at the detectors like raindrops on a skylight. To the rescue of HBT came a physicist who not only understood quantum statistics but who also understood the differences between the HBT and coincidence experiments.

Edward M. Purcell (1912–1997) was an American physicist who spent most of his career at Harvard. He won the Nobel Prize in 1952 for discovering nuclear magnetic resonance, the physical basis for MRI imaging. In the midst of the HBT controversy, Purcell wrote a letter to Nature[32] that explained the HBT results from the viewpoint of photons as wave packets. When the wave packets partially overlap, they display interference that causes excess fluctuations when observed in a square-law detector (a detector linear in intensity) like the photomultiplier tube. These excess fluctuations were precisely the positive correlations in the noise signals that HBT had observed at small path differences for the two detectors.

Purcell's wave-packet approach was analogous to the semiclassical approach that Twiss had originally taken, because classical electromagnetic waves from multiple sources have a spectral distribution that produces positive noise correlations at two detectors. But Purcell dug deeper, pointing out that photons are bosons—indistinguishable integer-spin quantum particles that obey Bose–Einstein statistics. Quite simply, the probability distribution for bosons is peaked at zero photons and decays monotonically with increasing photon number. In an attenuated beam of light, even if the average number of photons is around unity, it is still almost as likely to have two photons, or three or more in the apparatus at a time as it is to have one photon. In other words, photons tend to bunch together! This behavior for photons is known as "photon bunching" which, contrary to the view of the photon as a distinguishable particle, leads to positive correlation of photon arrival times in the two detectors. Furthermore, Purcell calculated that the coincidence rates in the coincidence-counting versions were too small to detect bunching. At the end of his letter, Purcell stated that "counting periods of the order of years would be needed to demonstrate the effect with the apparatus of Brannen and Ferguson. This only adds lustre to the notable achievement of Brown and Twiss."

HBT did not wait for the controversy to die down before taking the next step: stellar intensity interferometry. The lab experiment had shown that photon noise correlations were analogous to Michelson fringe visibility, so the stellar intensity interferometer was expected to work similarly to the Michelson stellar interferometer—but with better stability over much longer baselines because it did not need a reference. An additional advantage was the simple light-collecting requirements. Rather than needing a pair of massively expensive telescopes for high-resolution imaging, the intensity interferometer only needed to point two simple light collectors in a common direction. For this purpose, and to save money, Hanbury Brown selected two of the largest army-surplus anti-aircraft searchlights that he could find left over from the London Blitz. The lamps were removed and replaced with high-performance photomultipliers, and the units were installed on two train cars that could run along a railroad siding that crossed the Jodrell Bank grounds.

The target of the first test of the intensity interferometer was Sirius, the Dog Star near Orion. Sirius was chosen because it is the brightest star in the night sky, and it is the eighth closest star to Earth at 8.6 light-years and hence would be expected to have a relatively large angular size. The observations began at the start of winter in 1955, but the legendary English weather proved an obstacle. In addition to endless weeks of cloud cover, on many nights dew formed on the reflecting mirrors, making it necessary to install heaters. It took more than three months to make 60 operational attempts to accumulate a mere 18 hours of observations.[33] But it worked! The angular size of Sirius was measured for the first time. It subtended an angle of approximately 6 mas, which was well within the expected range for such a main sequence blue star. This angle is equivalent to observing a house on the moon from the Earth. No single optical telescope on Earth, or in Earth orbit, has that kind of resolution, even today.

Armed with the success with Sirius, HBT lobbied for the construction of a stellar interferometric observatory that could systematically measure the angular sizes of dozens of stars with a resolution that would be more than ten times better than the results of Michelson and Pease. Of course, the weather in Britain would never do. Fortunately, Twiss had recently taken a position in Australia that had favorable observing conditions, so the decision was made to build the observatory in Australia as a joint effort between the universities of Sydney and Manchester. A site was chosen outside the small town of Narrabri in New South Wales about 300 miles from Sydney. The weather was clear at Narrabri over 200 nights per year, and it was so far from civilization that the sky was perfectly black on moonless nights. The new stellar intensity interferometer was an ambitious project, composed of two large reflectors mounted on a 188-meter-diameter circular track, and the receivers were attached by two hanging cables to a central tower and control room, shown in Figure. 5.6. The entire installation took over a year to assemble, even after all the elements had been manufactured separately and shipped

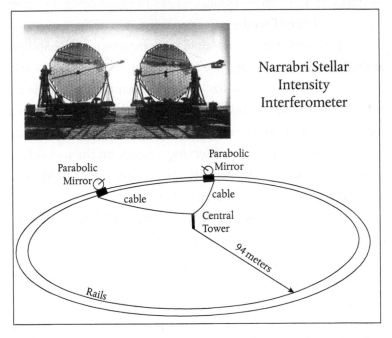

Figure 5.6 The Narrabri stellar intensity interferometer located in the outback of Australia operated from 1963 to 1972.

to Australia, overcoming numerous obstacles, ranging from import officials impounding the parts and demanding payment of tariffs, to steel wheels on the reflector chassis that were too soft and sagged under the weight. By mid-1963, almost 10 years after Hanbury Brown first conceived of stellar intensity interferometry, the Narrabri Observatory was ready for first light.

When they turned on the system and pointed it at their first selected target, β Centauri with a bright primary star and a dim companion, the results were all wrong. The observed correlations were far off the expectations. The team went over every mechanical and electronic detail of the system—Twiss even went back to his equations to see if somehow a mistake had been made—but nothing could reconcile their skewed measurements. With a sense of growing panic—years of preparation and large sums of research dollars were on the line—they trained the interferometer on the star Vega, and it worked perfectly! By sheer bad luck, when they had chosen β Centauri as their first test case, no one knew at that time that it was a triple star system. It was only after they made successful measurements of numerous other stars that they returned to β Centauri and showed that the bright primary was in fact two separate stars. This discovery was a potent demonstration of the power of stellar intensity interferometry.

Over the next eight years, the observatory logged 2,500 hours of observations and measured the angular diameters of 32 stars, completing their sky survey in February 1972. Prior to Narrabri, the angular size of only six stars had been known, measured by Michelson and Pease using *amplitude* interferometry. Narrabri, using *intensity* interferometry, brought the total number of known angular diameters to 38. They also were able to measure properties of binary stars like Spica, in the constellation Virgo. Spica is what is known as a spectroscopic binary—two companion stars that are too close to resolve, but by looking at Doppler shifts in spectral lines, the object is known to be composed of two stars orbiting each other at high speed. By combining the spectroscopic information with the interferometric information, the Narrabri interferometer was able to measure the actual distance to Spica at 84 parsecs. All previous distances to stars had relied on measurements of parallax.

The dramatic success of the HBT stellar intensity interferometer demonstrated not only the useful character of fluctuations in photon streams, but also the relative invariance of the technique to fluctuations in the atmosphere. However, even intensity interferometry has its limits, and atmospheric fluctuations do creep back in for weaker and more distant sources. There is no escaping the fact that Earth's life-sustain atmosphere keeps getting in the way of our efforts to see deeper into the night sky. But rather than trying to bypass the fluctuations, as HBT had done, another approach is to tackle the problem head on by measuring the fluctuations directly and subtracting them from the data. As challenging as that sounds, interferometry once again became the crucial element that made this work.

Adaptive Optics

You wouldn't know it, when looking up at the bright stars on a clear winter night, but the atmosphere above you is awash in violent agitation. Some indication of this violence can be seen in the deceptively peaceful twinkling of starlight. Take Sirius, for instance, on a cold winter night—the brightest star in the southern sky when it hangs above the horizon trailing Orion's heel. It's intense blue blazes and flickers in rapid random succession, heralding the invisible presence of quick density fluctuations driven by shifting temperatures and raging winds at increasing altitudes in the atmosphere. This mesmerizing twinkle, as beautiful and ethereal as it appears, was the greatest limitation faced by Michelson and Pease with their interferometric optical telescopes. It also motivated the launch of the Hubble and James Webb space telescopes, flying far beyond the limitations of the Earth's atmosphere. But space telescopes are exceedingly expensive, and Earth-bound telescopes can be made much bigger with much better resolution if the atmosphere were not there.

The fluctuations in the refractive index of parcels of atmosphere deflect rays and cause optical aberrations that distort telescopic images. The dynamically changing parcels have spatial and temporal scales that define what is known as astronomical "seeing." The characteristic size of consistent patches of light (speckles on the image plane of the telescope) is about $r_0 = 25$ cm where r_0 is known as the Fried parameter. Although the ideal angular resolution of a telescope may be given by λ/D, improving as the diameter gets larger, the angular resolution of an astronomical object is limited by the atmosphere to a value of λ/r_0—as if the telescope only had a diameter of 25 cm, even if the physical diameter were ten times or a hundred times larger. Therefore, for telescope diameters larger than about $2r_0$ (50 cm), the resolution becomes independent of the diameter of the telescope and is determined solely by atmospheric turbulence.

The turbulence-limited angular resolution for reasonably good seeing is about 1 arcsecond. This is equivalent to seeing a half-centimeter object at a distance of one kilometer (or a third-of-an-inch object at a distance of a mile). To give an idea how this affects astronomy, the angular size of Jupiter in the sky is about twenty times larger than the turbulence-limited resolution. This means that the best unaided telescopes on Earth, under good seeing conditions, can resolve Jupiter with about 300 pixels (or resolution elements). But Saturn can only be resolved with about 50 pixels and Uranus with only about 3 pixels.

Because of the transient nature of turbulence, there can be moments when a Fried "patch" temporarily grows to be the size of the telescope, and a sharp image can be seen for a short time, usually less than a tenth of a second, when the image is near the ideal resolution of the telescope. This is known as "lucky" seeing, and with the human eye, fine details of a target can be discerned through rare glimpses over time. When shifting from human observers to cameras, long-exposure photographic plates are limited by the Fried parameter, but short-exposure movies on digital cameras can still get "lucky" at certain moments. But hoping to get lucky is not a good way to do serious science.

At first look, turbulence seems to be a fundamental limit on astronomical seeing from the Earth. It is random and changes rapidly, producing complicated ray trajectories across the nearly 100 km thickness of Earth's atmospheric blanket. The distances involved, and the complex light patterns dancing across the face of the telescopes, present a serious challenge to imaging. Yet if the complex aberrations could somehow be tracked in time and compensated, then the wavefronts striking the telescope cameras could be made calm and smooth, and the full resolution of the large telescopes could be achieved. A first hint that this was possible came from a young astronomer in California in the 1950s.

Horace W. Babcock (1912–2003) was born in Pasadena, California, the son of the astronomer Harold Babcock who was on the staff of the Lick Observatory. After receiving his undergraduate degree at Caltech and his doctorate in astronomy at the University of California at Berkeley, the younger Babcock joined the staff at the Palomar Observatory in the foothills above San Diego. Babcock's scientific interest was in the magnetic fields of stars, and his main talent was designing and building instruments. Collaborating with his father, he invented an ingenious optical instrument that mapped solar magnetic fields, with which they discovered that the field of the sun's photosphere is organized in filaments.

As Babcock tinkered with the optomechanical design of his solar magnetograph, which used a fast optical scanning mechanism that switched polarizations at a rate above a hundred times per second, it struck him that such fast mechanisms might be able to track atmospheric turbulence that fluctuated at about the same rate. He envisioned a thin layer of oil on a conductive mirror whose thickness could be manipulated by voltages to cancel out the wavefront aberrations. The key to Babcock's insight was that a simple wavefront sensor, as simple as placing a knife-edge at the focal point of a lens, would be able to measure the shifts in the image to provide the control signal to the oil-film mirror. The closed-loop feedback between the wavefront sensor and the adaptive mirror would cancel the aberrations of the atmosphere. He published his idea in 1953 in a paper titled "The Possibility of Compensating Astronomical Seeing" in the *Publications of the Astronomical Society of the Pacific*.[34] Although the control of the oil film in the way he envisioned was not practical, it opened the possibility of zeroing out atmospheric turbulence to improve the astronomical seeing of large-aperture telescopes. However, astronomy was not the driving force for the first practical adaptive optics system. That came from an unexpected rocket launch from deep inside the Soviet Union.

Sputnik, the first artificial satellite, was launched on October 4, 1957 from a site in what is now southern Kazakhstan. The event created an existential crisis for the United States that forced it to reassess how it educated (or failed to educate) its citizens and trained them for high-tech jobs. Less known is that it launched dozens of high-tech half-secret start-up companies responding to business opportunities as well as dollars that began to flow from the military to meet the threat. One of these companies was the Itek Corporation, founded in Lexington, Massachusetts with an investment by Laurence Rockefeller, the son of John D. Rockefeller. In its first decade, it provided high-resolution cameras for US spy satellites flying over the Soviet Union. At the same time, a steady series of Soviet satellites of rising sophistication appeared in the night sky above the United States, and it became an urgent need to find them and to image them to asses any threat they may pose. But imaging satellites from Earth suffered from the same atmospheric turbulence as astronomical seeing.

In 1972, Itek's director of marketing, Richard Vyce, met with officials of the Advanced Research Projects Agency (ARPA) looking for business opportunities. ARPA was another entity that had been launched in 1957 in response to *Sputnik* as a way to generate new technology that could serve national security. Given Itek's expertise in imaging the ground from satellites, ARPA asked whether Itek could image satellites from the ground. Ryce returned to Itek and met with John Hardy, a British engineer who was widely regarded within the company as one of its most inventive and capable problem solvers. Hardy immersed himself in a study of the literature on imaging through atmospheric turbulence and found the paper by Babcock from 1953. Twenty years of optical technology development had elapsed by the time Hardy revisited the problem of pre-compensation of wavefront distortion, and Hardy thought of ways to improve both the wavefront sensor and the deformable mirror that might satisfy the requirements for imaging satellites. Ryce and Hardy returned to ARPA in late 1972 with a proposal, and ARPA issued a contract to Itek in the spring of 1973 to begin a proof-of-principle demonstration of the first adaptive optics system.

Hardy was put in charge of the project, and he enlisted two key Itek researchers onto his team: Julius Feinleib and James Wyant. Feinleib developed a deformable mirror based on the piezoelectric effect. This effect, discovered in 1880 by Pierre and Jacques Curie, causes a change in the thickness of a polar crystal when a voltage is applied to it. Feinleib fastened a thin mirror on top of a piezoelectric crystal that had a dense array of electrodes embedded in it. The voltage on each electrode could be controlled separately so that the vertical positions of each portion of the mirror above the electrode array could be adjusted independently. If the profile of the mirror across the distorted wavefront was adjusted to have the exact inverse surface relief to match the phase distortion, then an undistorted plane wave would appear as the reflected wavefront. To drive the electronics to match the wavefront required a phase-sensitive wavefront sensor. This was developed by Wyant using a white-light shearing interferometer, based on a Shack–Hartmann interferometer.[35] This is a multi-lens array placed one focal distance in front of an electronic camera, shown in Figure. 5.7. Each lenslet captures a local part of the

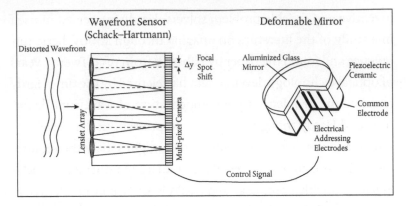

Figure 5.7 Two key elements of the adaptive optics system. The wavefront sensor is a Schack–Hartmann interferometer. The deformable mirror is a thin aluminized mirror on a multi-element piezoelectric stack.

wavefront and focuses it at different locations from the optic axis depending on the local distortion of the wave. By measuring the locations of all the different focal spots, it is possible to compute the wavefront distortion and to send this information to the deformable mirror.

Before the end of the year the three-person team had developed and performed a laboratory test of the first adaptive optics system using 21 piezoelectric actuators in a closed loop driven by the wavefront sensor. The technology was so new that the team was afraid the deformable mirror might vibrate itself to pieces under the rapidly updating high voltages it was receiving, but it held together and successfully compensated for designed wavefront distortions in a laser beam. The next step moved the system outdoors to a less controlled environment at the Air Force's Rome Air Development Center (RADC) located near Rome, New York. On June 4, 1974, the combined Itek and Rome teams succeeded in compensating wavefront distortions over a 300 m free-space range.[36] Hardy presented the results one month later at the Optical Society of America's Topical Meeting on Optical Propagation through Turbulence held in Boulder, Colorado.

The success of the Itek–Rome team, and the public disclosure of the results at Boulder, created a split that lasted for the next 20 years. ARPA's open support of the effort to demonstrate the feasibility of adaptive optics was now transferred to the Air Force where all further work was classified as top-secret. The military had significant resources and made rapid progress on ideas and performance. At the same time, academic groups, with fewer resources, began pursuing adaptive optics for astronomical uses (the conceptual design of an adaptive optical telescope is shown in Fig. 5.8) without knowing what the military was doing. For instance, in 1977, researchers at Bell Labs[37] and at Space Sciences at UC Berkeley[38] each made astronomical demonstrations with improved seeing of the star Sirius using adaptive optics, but the military was secretly much farther along. However, the academic/military split could not continue after a radical new idea swept into adaptive optics.

The first operational adaptive optical systems were mere demonstration tools that used much of the light of a star to drive the adaptive optics, leaving little light left over for imaging. This meant that only very bright objects, like Sirius, could be observed. As the optical systems became more efficient, lower magnitude stars could be observed, but still with limitations. A partial improvement came with the enlistment of natural guide stars—relatively bright stars that appeared in the same field of view as an object of interest. The adaptive optical system would use the nearby star to drive the feedback loop, enabling all the

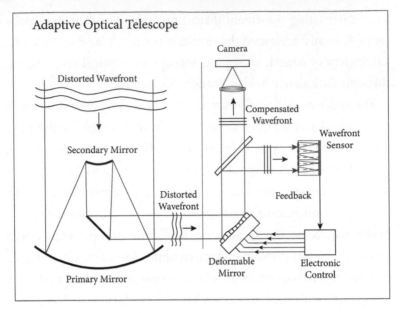

Figure 5.8 Adaptive optics system on a telescope. The distorted wavefront is reflected from a deformable mirror, where a part of the signal is input into a wavefront sensor that generates control signals to deform the mirror to cancel the distortions. The sensor is a Schack–Hartmann multi-lens system.

objects within the field of view of the telescope to be observed with compensated seeing. However, it was not always possible to find a star bright enough within the same field of view as an object of interest, making the use of adaptive optics hit-or-miss. What was needed was a technique that could provide an adaptive signal for any field of view at any magnification.

By 1981 the Air Force had begun experimenting with lasers to try to illuminate satellites when they were in Earth's shadow and invisible to the ground. Their key test site was a telescope installation on the Hawaiian island of Maui at the rim of the Haleakala volcanic crater. Hawaiian volcanos, with their high altitude and low humidity, make exceptionally good sites for astronomical seeing. Julius Feinleib, who had developed the micromirror for Itek, by this time had formed his own company and was doing contract work at the Haleakala telescope. One night, in the summer of 1981, he was standing outside the telescope housing looking up into the night sky when the high-power green laser fired. From his low angle of view relative to the direction of the laser, he noticed a bright green spot at the vanishing point of the laser beam. This optical effect was caused by light backscattering from density fluctuations in the air (for the same reason that the sky is blue), in a process known as Rayleigh scattering. It immediately occurred to him that Rayleigh backscattering, if viewed by the telescope, could be used as an artificial guide star for the adaptive optical system to compensate the atmospheric turbulence.

Feinleib began lobbying what had become Defense ARPA (DARPA) to fund his laser guide star idea. The program managers were receptive but unsure if the approach would work, so they brought together a top-secret team of experts, known as the Jasons, to assess the idea. The Jasons were military advisors, founded in 1958, yet another offshoot of *Sputnik*, as a team of the country's top technical experts. They were the reality check on new ideas that may be crack-pot and destined for the trash can— or may be ingenious and deserving of funding. In the summer of 1982, the Jasons convened in San Diego to consider the merits of Feinleib's Rayleigh guide star idea, and they decided that it was workable and should be tried.[39] But in the process, a new idea emerged.

One of the attendees at the San Diego meeting was the physicist Will Happer from Princeton. He had only joined the meeting on a whim at the urging of a friend who thought he would find it interesting. During the discussions about the Rayleigh guide star, which produced backscattered photons mainly from low in the atmosphere, atmospheric imaging experts like D. L. Fried (of the eponymous Fried parameter) pointed out that the low altitudes would limit the angular sizes that could be compensated and hence would limit the size of the telescopes for which the Rayleigh guide star could help. As Happer listened and learned for the first time about imaging through atmospheric turbulence, he had a sudden thought. As a spectroscopist who had worked on fluorescence of alkali metals, like sodium, he had learned somewhere that there was a layer in the upper atmosphere that contained a dilute gas of sodium atoms. He realized that if this high-level layer could be excited by a green laser, then the sodium ions would fluoresce with yellow light and act as a distant artificial star that could work almost as well as a natural guide star. Furthermore, the fluorescence could be much brighter than the Rayleigh backscatter, reducing the power needed for the laser. This alternate approach to the laser guide star is called the "sodium guide star."

Therefore, there were two outcomes of the 1982 Jason meeting for creating artificial guide stars for adaptive optics: Rayleigh guide stars and sodium guide stars, with pros and cons for each. The Rayleigh approach could use a laser of any wavelength, while the sodium guide star required lasers of a specific color. On the other hand, the sodium guide star was much higher in the atmosphere than the Rayleigh guide star and could compensate the atmosphere on much larger telescopes. Faced with these two alternatives and their trade-offs, DARPA decided to fund both. The Rayleigh approach was pursued at the Starfire Optical Range at Kirtland Air Force Base outside Albuquerque, New Mexico, while the sodium guide star was pursued by Lincoln Labs of MIT and tested at White Sands, also in New Mexico. Both efforts were classified top-secret. At that time, the US military was only interested in imaging enemy satellites, but there were more universal needs for adaptive optics.

In 1984 the European Southern Observatory (ESO), that operated the world's largest astronomy observatories, realized that the Hubble Space Telescope, then under development by NASA, would have limited resolution despite its enviable position above atmospheric turbulence. The ESO committed to building the most ambitious telescope on Earth called, prosaically, the Very Large Telescope (VLT). It would consist of four separate telescopes, each with a primary mirror of 8.2 m diameter, and four movable auxiliary telescopes of 1.8 m diameter. When used together as an interferometer, the Very Large Telescope Interferometer (VLTI) would have an effective baseline up to an unprecedented 130 m. However, such a large Earth-bound telescope could only benefit from its large diameters and baselines if it could compensate atmospheric turbulence. Therefore, at the same time the ESO was committing to building the VLT, it was also committing to adaptive optics, and a French astronomer was soon paralleling the US military efforts, helping to bring the academic and military schism to a head.

Antoine Labeyrie (1943–) is a French astronomer who graduated in 1965 from the Grand École Supérieure d'Optique, known as SupOptique, which is the leading technical university in France for optiCaltechnology. His doctoral studies on astronomical telescopes focused on the problem of atmospheric seeing, and he began developing the first long-baseline optical interferometer.

It used two optical telescopes separated by 13.8 m[40] in the Fizeau configuration where an image of the target is formed on the electronic camera modulated by the interference fringes set by the baseline length. First fringes in the dual-telescope system were observed on August 13, 1974, demonstrating the first long-baseline optical stellar interferometer.[41] The telescope pair was then moved to an observatory on the Plateau de Calern in the coastal highlands to the west of Nice, where it operated for over a decade with an extended baseline of 35 m, measuring angular diameters and effective stellar temperatures.

With his experience developing this first optical long-baseline stellar interferometer, Labeyrie began thinking about direct adaptive compensation of atmospheric fluctuations. Given his expertise in coherent imaging, it was natural for him to consider using the best source of coherence—the laser—to solve the guide star problem. In 1985, with his colleague Renaud Foy, Labeyrie published a paper outlining the details of both Rayleigh backscattering and sodium fluorescence as artificial guide stars in adaptive optics systems—precisely the approaches that the Jasons had recommended in secret to the US military. This paper would have been a banner publication for Labeyrie and Foy if the Jasons had not gotten there first. Worse, some of the Jasons later accused Labeyrie and Foy of having seen a classified report that the Jasons had circulated in 1984 to its members. When later asked about it, Foy vehemently denied any help from the Americans.[42]

This controversy brings to mind the question of whether Michelson knew of Fizeau's proposal for stellar interferometry. Michelson had all the talent and tools needed to invent stellar interferometry on his own. If a simple word were passed in a hallway during his visit to the Institute d'Optique about measuring the stars, then Michelson would have had everything he needed to develop stellar interferometry. Once an idea is in the air, someone with keen insight can logically follow it to its conclusion. It is no accident that the history of science has so many examples of discoveries that are made by different groups within months of each other. Sometimes an idea just has its time.

The laser guide star controversy also highlighted the essential tension between the military and the academic threads of adaptive optics research. The military had much better resources than the academics and was far ahead, as the academics re-invented the wheel and implemented it with inferior equipment without knowing it. The US National Science Foundation (NSF) began receiving grant proposals on adaptive optics that were advanced for academic efforts but were already obsolete relative to the military efforts. This situation was rightly viewed as unhealthy, and the NSF filed a request to the military in 1992 to declassify adaptive optics research. The military surprisingly agreed, and adaptive optics research was declassified by 1994. When the academics saw what the military had accomplished, they were awed, but some were a little angry at their wasted efforts. However, they now had access to the best studies and equipment, and it changed the history of large Earth-based telescopes in astronomy and their uses—like looking for second Earths orbiting nearby stars.

6

Across the Universe

Exoplanets, Black Holes, and Gravitational Waves

... we had every advantage we could desire in observing the
whole of the Passage of the planet Venus over the Sun's Disk.[1]

Captain James Cook (1769)

The universe holds such marvels that our minds boggle at
their magnificence and strangeness. Yet these marvels are
so remote that we yearn to bring them closer to explore their
mysteries in finer detail. In our quest to bridge astronomically
large distances, we turn to the magic of interferometry that
divides the tiny length-scale of light—a wavelength λ about a
micron in size—by the even smaller angular size α of a distant
object to yield an intermediate size $d = \lambda/\alpha$ that we can handle
and measure on Earth. Although astronomical distances can
be incomprehensibly large and beyond reach, the characteristic
length d can be terrestrial in size. Given the range of angular
sizes in the sky, the characteristic length d can be the size of the
primary mirror of a telescope, or the distance between two or
more telescopes separated by baselines of meters or kilometers
and even by continental distances. To gain a semiquantitative
intuition about this, consider two telescopes on a baseline d =
50 m (about half a football field) observing in the near infrared

Interference. David D. Nolte, Oxford University Press. © David D. Nolte (2023).
DOI: 10.1093/oso/9780192869760.003.0006

with a wavelength $\lambda = 1$ micron. This can resolve an angular size of about 4 milliarcseconds (mas), which is the angular size of our sun seen from 7 light-years away. This modest telescope would be sufficient to detect planets around stars hundreds of light-years away if the glare of the central star can be removed.

The Search for New Worlds

Ever since we discovered that our Milky Way is an island of stars in a vastly larger universe, mankind has been compelled to ask whether, out among the stars, there might be other Earths supporting life. For a long time, there was no evidence that planets outside our solar system even existed, and some believed that we were alone in the universe. However, it is now estimated that there may be as many as four billion Earth-like planets in our Milky Way Galaxy. Most are far too far away ever to visit, even by unmanned craft. But within 50 light-years from Earth there could be as many as 300 Earth-like planets. Fifty light-years may be a great distance, but star-drive technology capable of transporting unmanned missions over such distances could well be developed within this century. It would take about 200 years for the craft to get to those outer stretches, and another 50 years to transmit information back, but it would allow us to "explore" such distant planets with a quarter-of-a-millennium turnaround time—if Earth society is willing to spend the time and money to do it. But first we have to find those planets.

As of the writing of this book (2022) there are more than 4,000 confirmed extra-solar planets, known as exoplanets. The fact that I have to quote the publication date says something about how fast that number is increasing. The first discovery of an exoplanet orbiting a main-sequence star was made nearly 30 years ago in 1995 by Michel Mayor and Didier Queloz of the University of Geneva. They used extremely precise astrometric measurements of the wobble of the star 51 Pegasi caused by the tug of a Jupiter-sized planet orbiting around it. The wobble was detected using the Doppler shift of the frequency of light emitted by the star, a shift that was only one part in ten million, corresponding to a change in speed of the star toward and then away from the Earth by only 50 m per second. Measurable Doppler shifts such as this are only possible for very large planets orbiting very close to their star, so these are not the long-sought exo-Earths.

An alternative to Doppler detection of exoplanets is the simple transit method: detecting the trace of a distant planet across the disc of its star. For instance, the famous first voyage of Captain Cook to Tahiti was to observe the transit of Venus across the face of the sun—and along the way he ran into and claimed both New Zealand and Australia for Britain. During a transit, the light of the star dims slightly when part of the star's disc is occluded by the passing planet. This became the dominant method to detect exoplanets with the launch of the Kepler Telescope in 2009, and most of the known exoplanets were detected by it and by its successor TESS that was launched in 2018. However, detection of transits still requires relatively large planets orbiting relatively dim stars, such as brown dwarfs, which are not hospitable hosts for stable planetary environments. Furthermore, transits only occur for planetary orbits that are perfectly aligned so that the planet passes between us and the star. Few orbits have this unlikely configuration. The fact that we have detected so many exoplanets this way says something about how multitudinous exoplanets are. Still, only about 25 habitable candidates (out of roughly 300 to be expected) have been identified to date within 50 light-years from Earth.[2] Therefore, to be sure we don't "miss" a prime candidate requires a different technology, one that can detect *all* Earth-like exoplanets within a 50 light-year radius, regardless of their orbital inclinations. This technology could be *nulling stellar interferometers.*

At its simplest, nulling adds two signals together that are perfectly out-of-phase to create destructive interference. A nulling stellar interferometer acquires two images of a distant solar system from two telescopes separated on a long baseline, and then adds the images out-of-phase to cancel the light from the star. In practice, this puts the image of the star at a destructive interference fringe on the image plane, as shown in Figure. 6.1. The fringes are not directly visible in the image—they simply denote where signals would add destructively versus constructively. The planet, which is off the optic axis, may fall on a constructive fringe, making it visible because the glare of the star has been removed by a destructive fringe. Depending on its orientation in the sky and on the direction and length of the baseline, the planet may also fall on a destructive fringe. Therefore, for this approach to work, the baseline needs to be adjusted to move the fringes. To solve this problem, Ronald Bracewell at Stanford proposed in 1978 a space-based pair of telescopes that would rotate around a central axis, causing the interference fringes to rotate. As the exoplanet passes through constructive and destructive fringes during the rotation, the intensity signal would be modulated at a frequency that is twice the rotation frequency, allowing accurate measurements of the planet's light while subtracting out the stellar background. This became known as the Bracewell interferometer.

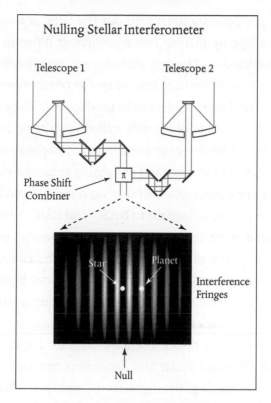

Figure 6.1 A nulling stellar interferometer in the Fizeau (imaging) configuration. The images of a star from two identical telescopes are added 180° out of phase to cancel the light of the star at a dark interference fringe, while the image of the planet falls on a bright fringe. The fringes are not directly observable—they serve only to show what points on the image have destructive or constructive interference.

In the early 2000s, nulling interferometers were the next hot items to be put into space. Both NASA and the European Space Agency (ESA) competed to develop stellar interferometer satellites that would go into deep space to the sun–Earth Lagrange point L2. These had the mission names Terrestrial Planet Finder (TPF) and Darwin, respectively. However, an alternative non-interferometric approach was also being pursued known as a coronagraph. Coronagraphs zero-out the light of a star using a single telescope by placing two apertures at different locations inside the telescope. The first coronagraph was invented by the French astronomer Bernard Lyot in 1931 to observe prominences on the sun. Similar principles can be used to search for exoplanets, and as the first decade of the new millennium came to a close, a heated competition arose between the interferometry and coronagraph groups. The debates and in-fighting apparently became so intense that the exoplanet astronomy community could not agree on a clear mandate to present to NASA and ESA.[3] Subsequently, both projects were dropped by the space agencies in favor of continued support of the transit method. The cancellation of the space-borne interferometers came as a bitter blow to those groups, but the quest to find second Earths using interferometry eventually came back down to Earth.

By using adaptive optics to compensate for atmospheric turbulence, Earth-based stellar interferometers can look for second Earths by combining the light from multiple telescopes on adjustable baselines. This shift away from space platforms opens new possibilities for the direct detection of exoplanets. Earth-based telescopes can be extremely large, thanks to adaptive optics and the Schack–Hartmann interferometers that drive them, giving them much higher spatial resolution than is possible with the necessarily smaller diameters of the space-borne telescopes. Furthermore, planets emit a wide range of frequencies, not all optical, and are natural sources of infrared light, while stars are less bright in the infrared, reducing the amount of direct glare that needs to be canceled by nulling interferometry to see a planet. For these reasons, the Earth-based interferometers look for planets in the infrared. On the other hand, infrared light suffers from absorption by water vapor in the atmosphere. Therefore, to take advantage of large telescopes in the infrared using adaptive optics, it is necessary to find a prime location that has still skies free of moisture.

The high desert of Atacama on the Pacific coast of northern Chile is a good place for star watching because it is one of the driest places on Earth. The winds off the cold Humboldt Current in the Pacific lose their moisture as they rise to cross the coastal range before plunging into the Atacama gap before the Andes. Some weather stations in the Atacama have never seen a drop of rain, and remnant riverbeds throughout the area have been dry since the last ice age. The Atacama has a barren beauty of stark rust-colored mountains on top of which are situated some of the world's most powerful telescopes that take advantage of the still and dry desert air to look into the night sky. These installations include ALMA, the Atacama Large Millimeter Array, the largest single-site astronomical system, having baselines up to 16 km long and achieving angular resolutions down to 7 mas. It images the deepest depths of the universe using the far infrared and microwave frequencies. Several kilometers away from ALMA is the VLT, the Very Large Telescope, which works at visible and infrared frequencies. The VLT is located on the flattened top of an 8,000-foot peak called Cerro Paranal. What makes the VLT so *very* large is that it consists of four separate large telescopes that can be combined into a single interferometric imaging platform using multiple baselines up to 130 m. This is called the VLT *interferometer*, or the VLTI.

The VLTI is the world's largest and most complex optical and infrared telescope array, shown in Figure. 6.2. It has four fixed large unit telescopes (UTs) with 8.2 m diameters complemented

Figure 6.2 The Very Large Telescope Interferometer (VLTI) on the peak of Cerros Paranal in the high Atacama Desert in Chile. The optical and near-infrared facility has four large telescopes and multiple smaller telescopes connecting multiple interferometric baselines up to 130 meters.

Credit: J. L. Dauvergne & G. Hüdepohl (atacamaphoto.com)/ESO.
https://www.eso.org/public/images/eso-paranal-51/

with four movable auxiliary 1.2 m telescopes within a 200 m radius footprint. The VLT saw first light in 1998. The first implementation of adaptive optics was in 2001 followed closely by the first interferometric observations. Since then, both the adaptive optics capabilities and its long-baseline interferometry have received steady improvements by the installation of laser guide stars, increasing the number of actuators on the deformable mirrors, and combining more of the telescopes into its interferometric mode. In 2016 the complete instrument, called GRAVITY VLTI, saw first light, combining the light of all four UTs. This set the stage for a landmark in the history of interferometric astronomy.

A star in the constellation of Pegasus has the non-descript name of HR 8799. It is a 30 million-year-old main-sequence star about 130 light-years from Earth. Its claim to fame was the discovery in 2008, using adaptive optics, that three giant planets were orbiting the star, and a fourth planet was discovered a year later. All of these planets are nearly ten times larger than Jupiter. The fourth planet, named HR 8799e, orbits its star with a period of 45 years (compared with 12 years for Jupiter), giving it a large orbital radius. This made it a prime target for the VLTI GRAVITY instrument. On the night of August 28, 2018, the GRAVITY instrument combined the light from all four of the large telescopes trained on HR 8799. Although the star was ten thousand times brighter than the planet, the light was sufficiently nulled in the interferometer to allow the light of the planet to be seen. Even then, the planet was so dim it took an exposure time of 100 seconds to record it.[4] Despite the advantages of the adaptive optics and long baselines, the authors felt compelled to state in the paper, published in May 2019, that the "seeing conditions were average to good"—astronomers today are still hostages to the quality of the atmospheric seeing. This observation of HR 8779e was the first interferometric observation of an exoplanet.

New paradigms usually start slowly and then accelerate. It took a year after Queloz and Mayor discovered the first exoplanet in 1995 for a second exoplanet to be reported.[5] Three years later a whole planetary system was discovered around Upsilon Andromedae.[6] Within a decade of that, literally hundreds and then thousands of exoplanets were being discovered by the Kepler satellite. But all of these were gas giants, and most were extremely close to their stars. Even HR 8799e, the first to be detected using interferometry, is a gas giant and not the kind of rocky planet that could support life, and it had already been observed using other means, so it was not a discovery, only a demonstration. Yet the promise of interferometric discovery of rocky exoplanets in the habitable zones of nearby stable stars remains strong. The VLTI will continue operating for at least the next two decades even as the extremely large telescope (ELT), a 39 m diameter segmented-mirror telescope, comes online on the flattened peak of Cerros Amazones about 15 km from the VLT. The ELT will look even deeper into the cosmos, but the VLTI remains the leader in looking closer to home by using interferometry. Eventually, when formations of satellites start flying far from Earth, nearly all of the several hundred rocky planets within a 50 light-year radius will come into focus, and we may consider launching a deep-space probe to explore a second Earth. But before then, interferometric astrophysics has many more roles to play and many more objects to image, like the monsters that lurk at the heart of most galaxies, even our own.

Seeing Black Holes

Supermassive black holes—black holes with masses greater than a million suns—are among the great mysteries of the universe. Although they are relatively common, occurring at the center of most (if not all) galaxies, by some arguments they should not exist at all. Supermassive black holes are fundamentally different than "ordinary" black holes that form as the result of the gravitational collapse of stars a bit larger than the sun. In our own Milky Way

Galaxy, there may be millions of stellar black holes floating about unseen. At the center of our galaxy, 26,000 light-years away, is a single supermassive black hole in a region of the constellation Sagittarius designated as Sgr A*. By using adaptive optics, astronomers have resolved the orbits of individual stars circling Sgr A*, for which Reinhard Genzel and Andrea Ghez received the Nobel Prize in Physics in 2020. From the closest approach of one of the stars in its orbit, the object is known to have a radius that is less than 6 light-hours, which is smaller than the orbital radius of Neptune. Yet from the period of the orbits, the central mass is known to have about 4.3 million solar masses. These data provide clear evidence that it is a supermassive black hole. But seeing is believing, and there was a race to image it directly. Yet why is it there?

Thirteen billion years ago, as the Milky Way Galaxy condensed out of the early universe, an intermediate-sized black hole at its core may have begun to accrete mass from surrounding gas clouds and stars and may have merged with other smaller black holes, growing in time to the million-solar-mass object it is today. Yet by galactic standards, our own supermassive black hole is relatively small and is at the lower end of the supermassive spectrum. By looking out into the universe, and far back into time, there are galaxies harboring ultra-massive behemoths at their cores containing billions of solar masses. Most of these are associated with ultra-energetic quasars. A quasar has a luminosity that outshines the light from all the other stars in its galaxy by a hundred fold. The age of these galaxies is only about a billion years after the Big Bang, which is too little time for such supermassive black holes to have formed by simply accreting mass. Therefore, numerous astrophysical theories have been proposed to explain how such large sizes formed in such short times, including theories for the direct collapse of non-rotating gas clouds into black holes without first forming stars. There are more speculative ideas that propose that such black holes formed during the Big Bang, or even (if there was such a thing) *before* the Big Bang, perhaps aided by dark matter.

In our Milky Way's own back yard, a mere 55 million light-years away, is the galaxy that was the eighty-seventh entry in the list made by Charles Messier in 1781 when he was cataloging all the brightest nebulae in the sky. The M87 galaxy is a monster with about one hundred times more stars than our Milky Way. As an elliptical galaxy lacking any beautiful spiral-arm features, it may have grown from the collision of a hundred smaller galaxies. As these galaxies merged, their central black holes may have merged too, creating a supermassive black hole of 6.5 billion suns. The M87 supermassive black hole is one of the largest confirmed black holes, and it is also one of the closest, which raised the tantalizing possibility that it could be imaged directly from Earth, if only the Earth had an interferometer big enough to resolve it.

There is a logical limit to the idea of long-baseline interferometry. If the resolving power is proportional to the width of the telescope, then there is an obvious limit to the size of Earth-bound telescopes—the size of the Earth! Imagine, then, building a telescope whose diameter is equal to the diameter of our planet. What could it resolve? Going to the formula for the resolution of interferometry and plugging in the radius of the Earth and picking a radio wavelength of 1 mm, the smallest resolvable angle is 0.1 nano-radians or 20 micro-arcseconds (0.02 mas). This would be like resolving a baseball on the surface of the moon or reading a newspaper in Paris while looking from New York. It would also be equivalent to imaging an object the size of our solar system from 55 million light-years away. Therefore, to image the M87 black hole directly would require a baseline that was an appreciable fraction of the size of the Earth. Such an instrument did not exist in 1999 when it was first proposed, nor even 10 years later in 2009 when Shep Doeleman (1967–) from the Harvard–Smithsonian Center for Astrophysics began organizing an international collaboration to make it happen. But the history of very-long-baseline interferometry (VLBI) astronomy at that time boasted decades of rapid technological developments that lent optimism in the project. The key was to find a way to fix the phases among multiple receivers without using any local reference wave.

When Hanbury Brown was at Jodrell Bank using intensity interferometry to make the first measurements of the angular sizes of radio sources, one of his students was Roger Jennison who had enrolled in Manchester in 1948 after coming out of the military radar services. By the time Hanbury Brown began planning the Narrabri optical stellar interferometer, Jennison had obtained his PhD and was lecturing at Manchester while continuing work on radio astronomy at Jodrell Bank using radio-wavelength intensity interferometry. As with the optical intensity interferometer, the radio-wave version of intensity interferometry was limited by the brightness of the sources that could be studied, and Jennison began to ponder how to return to phase-sensitive amplitude interferometry over long baselines. At that time, Martin Ryle, the master of radio interferometry who would receive the Nobel Prize in 1974, believed that there was a fundamental limit to the distance that two radio antennas could be linked to do interferometry because of the random phases caused by the ionosphere—limited by the analog of the Fried parameter scaled to radio wavelengths. At that time, the limiting link distance was only several thousand wavelengths or about 1 km.[7]

In a 1976 radio interview of Roger Jennison, hosted by the National Radio Astronomy Observatory, he gave a lively description of how he began to tackle the problem of long baselines in radio interferometry. The multiple-element interferometers in the mid-1950s worked on pair-wise baselines between two receivers at a time. The phases were not measured directly, rather the pair-wise phases were contained in a mathematical product of two complex visibilities multiplied together. The result was a complex phase that depended on the phase errors between the two receivers, and when these got too large, they ruined the interferometry. Jennison then tried something surprisingly simple, and he was amazed at the result. Rather than only working with two receivers at a time, he took the complex visibilities of three baselines among three receivers and multiplied them together (taking the complex conjugate of one of them). As if by magic, the three separate phase errors vanished in the resulting product! Jennison published his idea in 1958, and in the 1976 interview he said how remarkable and magical he still found it to be. Jennison's approach is called "closure phase," and it is what makes long-baseline interferometry possible when there are direct links among the receivers. But what if the receivers are so far away that these communication links themselves suffer from phase errors? Even the closure phase gets lost. But just as closure phase was a surprisingly simple solution to the problem of long baselines, an equally simple solution exists for very long baselines—magnetic tape synchronized to atomic clocks.

The idea of using magnetic tape to record astronomic radio signals came first to Leonid Matveyenko at the Russian Lebedev

Physical Institute in Moscow in 1962, but Western scientists were unaware of his idea until a visit to Russia in 1963 by the director of Jodrell Bank. The scientists attempted to form a collaboration to test the idea, but the red tape of the Cold War made it impossible. Meanwhile, an important advance was made by mixing the raw signal with a stable local oscillator to produce an intermediate-frequency signal that was much easier to record on magnetic tape. Around the same time, rubidium and hydrogen masers became available, providing time accuracy down to one part in fourteen orders of magnitude. At this point, the approach appeared easy— at each antenna, regardless how far apart they are, record the intermediate-frequency radio signal on the magnetic tape referenced to the time provided by the maser. Then the tapes can be stored and transported to a central facility when convenient, and the interference fringes could be reconstructed through analog or digital means at any later date.

The first attempt to do this was in 1965 by a collaboration between Cornell University and the National Radio Astronomy Observatory (NRAO) that ran the Arecibo radio telescope in Puerto Rico. At the same time a Canadian group launched a similar project, and a friendly competition arose to see who would succeed first. The Cornell–NRAO team tried to find reconstructed fringes from data taken in January and February of 1967 over a baseline of 2,557 km spanning 5 million wavelengths at 49 cm (610 MHz), but no fringes emerged from the data. This failure opened the door for the Canadians who were successful, but only over a very modest distance of a mere 200 m. Nonetheless, it demonstrated the feasibility of remote recording and reconstruction, which spurred on the Americans who achieved interference fringes over a 220 km baseline on May 8, representing the first true VLBI demonstration. The Canadians subsequently demonstrated fringe reconstruction over a baseline of 3,000 km on May 21, 1967.

During the next five years, magnetic recording was combined with Jennison's closure phase to produce VLBI systems that began high-resolution observations of astronomical radio sources, becoming a common approach by 1974. The hydrogen masers that were needed to run the systems were transported to the distant sites by flying them on seats in first class on commercial airlines because of arcane transport regulations at that time.[8] Once the problem of long baselines had been solved (extending to 10,000 km by adding a radio site in Australia), the next step to improving angular resolution of the sky required shortening the wavelengths from centimeters to millimeters. This was accomplished through the 1980s, culminating in the Very Large Baseline Array (VLBA) anchored in Socorro, New Mexico operating at millimeter wavelengths with a maximum baseline of 8,000 km.

Despite the existence of more than a dozen millimeter-wavelength radio telescopes spread across the world at the turn of the millennium, they all had various incompatibilities that prevented them from being combined into a single coherent instrument. In 2009, Shep Doeleman from the Harvard–Smithsonian Observatory organized the Event Horizon Telescope (EHT) project to put all the pieces together. The principal task was to remove the various incompatibilities so that each interferometer could be tuned to the same wavelength, synchronized by local atomic clocks, and all pointed at a single target on the same nights with good astronomical seeing conditions at all the locations all at the same time. Eight telescopes were selected to be part of the first-generation EHT, shown on the map in Figure. 6.3.

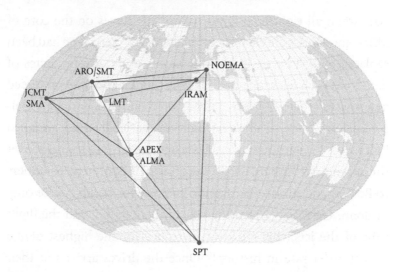

Figure 6.3 Locations of the telescopes of the Event Horizon Telescope (EHT).

Two are from the Atacama Desert in Chile, two from Mauna Kea in Hawaii, and single telescopes in Arizona, Spain, and Mexico. The eighth, used for calibration purposes on one of the longest baselines, is a telescope at the South Pole. An operation wavelength of 1.3 mm (230 GHz) was chosen at which all of the eight telescopes could operate. At this wavelength, and on these baselines, their estimates suggested that several supermassive black holes could be resolved, including M87, as well as the black hole in our own galaxy.

Preparations took seven years before the EHT saw first light. Most of that time was spent on electronics and software as the atomic clocks were integrated into each of the telescopes. In late 2016, when the EHT began operating, the final biggest challenge was finding a moment when the astronomical seeing was good at all locations. This finally happened on four nights in April of

2017 when all eight telescopes trained their eyes on the core of M87—the skies were clear for all of them. Magnetic tape had been replaced by solid-state drives that stored about 350 terabytes of data from each telescope per day. The four clear nights produced 5 petabytes (5×10^{15} bytes) of data stored on hundreds of specialized helium-filled terabyte hard drives that collectively weighed half a ton.[9] This amount of data is much too large to transmit over the internet, so the hard drives were flown by commercial jetliners to Boston and to the Max Planck Institute for Radio Astronomy in Bonn, Germany. Given the size of the data set, and the flight time of the jets, this set a world record for the highest digital data transfer rate in history.[10] Once the drives arrived at their respective locations, they were mounted into scores of hard-drive racks, and the data crunching began.

Teasing images out of interferometric data is a monumentally difficult job. To the eye, graphs of the data from each telescope look no different than graphs of random noise. To find the parts of the signal that carry image information requires extremely accurate knowledge of the absolute movement of each telescope through three-dimensional space as the Earth spins on its axis and orbits the sun. These complicated three-dimensional trajectories then need to be synchronized to the respective atomic clocks. By applying these exactly known times and locations to the data, and calculating all the closure phases, and shifting the data by exactly calculated amounts, and then adding the signals from the different telescopes—only then do interference fringes appear in the superposition of signals. But at this point, all that has been achieved is a periodic modulation envelope on the seemingly random noise. This still has to be put through complex calculations to generate an image.

Because the signal processing and image generation are such a difficult tasks, with many different algorithms to select from and many choices to make, in June 2018 the EHT data analysis team divided itself into four separate image-processing groups, all with access to the full data set, but all working completely independently, making their own individual choices with no communication allowed among the four groups. They worked for seven weeks in isolation, each wondering what type of image the other groups were seeing. Then on July 24 they reconvened in a room on the MIT campus in Cambridge, Massachusetts, to compare results. As each group put up their best image, Katherine Bouman, who is an EHT working-group coordinator for image processing, described that "everyone immediately started clapping and laughing,"[11] because all four images, though slightly different, showed the same broad features.

What they saw was a dark disk surrounded by a bright ring of light, brighter on the bottom than on the top, shown in Figure. 6.4. The dark disk matched theoretical computer simulations of the shadow cast by the black hole illuminated from its far side by its bright accretion disk and bent by gravity around the black hole. The asymmetry in the ring brightness also matched theoretical simulations of a rotating accretion disk, as the part moving toward the Earth is given a Doppler boost, making it brighter than the part moving away from us. Despite the first glimpse of success on that day in Cambridge, the imaging team continued to test and retest their algorithms against all possible images that could fool them. Finally, after an additional nine months of fine tuning, and writing a total of six simultaneous papers for publication, the EHT made dual simultaneous announcements: one at the National Science Foundation (NSF)

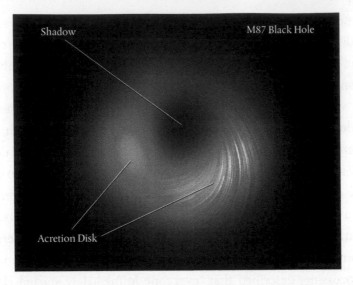

Shadow

M87 Black Hole

Acretion Disk

Figure 6.4 Scientists obtained the first image of a black hole, using Event Horizon Telescope observations of the center of the galaxy M87. The image shows a bright ring formed as light bends in the intense gravity around a black hole that is 6.5 billion times more massive than the sun.

Credit: Event Horizon Telescope Collaboration.
https://apod.nasa.gov/apod/ap210331.html

in Washington, DC and the other at the Max Planck Institute in Bonn, Germany. By using the exquisite sensitivity of interferometry, the EHT team had succeeded in making the first image of a black hole—55 million light-years away with an event horizon as big as our solar system.

With the success on M87, the EHT turned its arrays of telescopes on Sgr A* at the center of our own galaxy. Although it is a thousand times closer, it is also about a thousand times smaller, making its angular size roughly the same as M87. However, there is a lot of star dust between us and Sgr A*, creating new challenges for the data acquisition and the imaging algorithms. During the preparation of this book for publication, the EHT released its first image of Sgr A*, and there are many other potential targets for the EHT. The imaging of M87 and Sgr A* represent the start of a new era in ultra-high-angular-resolution astronomy. These first images of black holes may garner a future Nobel Prize, just as the first detection of gravitational waves from the merger of two black holes in 2015 was awarded a Nobel Prize in 2017 for the leaders of the team. That feat also was accomplished using the power of interferometry—not to see, but to listen.

Listening to Gravity

The Italian scientist Marco Drago began his 15 minutes of fame when a simple email alert appeared on his computer in late 2015 at the Max Planck Institute for Gravitational Physics in Hannover, Germany. His email alert was set up to review selected signals coming from the kilometer-sized LIGO (the Laser Interferometric Gravitational Observatory) experiment in the United States.[12] Around noon his computer opened an alert panel on his display. When he pulled up the LIGO signal, expecting to see the usual noise, he saw instead something that must have made him pause. It was a classic transient signal that looked like a numerical simulation. In fact, his first thought was that someone had injected a simulation into the LIGO data system, either as a joke or as a test. He checked with colleagues to see if that was the case, and they called the operators in the US to ask whether any synthetic data had been injected, but the operators said no testing had been done for that night. With a growing sense of anticipation, he wrote an email to the other scientists on the collaboration to verify what he was seeing, cc-ing the email to the entire mailing list of the collaboration. By the time the sun was rising over the Eastern United States, the LIGO servers were being flooded by a flurry of emails alerting the team leaders that a highly unlikely, yet eagerly anticipated, discovery may have been made—and Marco Drago had chanced to be the first human to put eyes on gravitational waves predicted by Albert Einstein a hundred years earlier.

Einstein, working in Berlin in 1916 at the height of WWI, was exploring the consequences of his newly developed theory of general relativity, describing the physics of gravitation, when he discovered that a wave equation emerged, predicting the propagation of gravitational waves (GWs) traveling at the speed of light.[13] When he did a back-of-the-envelope calculation to see how strong such a wave could be, he decided that it was far too weak ever to be detected, and he turned his back on the idea after he published it in a 1916 volume of the *Proceedings of the Royal Prussian Academy of Sciences*. Fifty years later, Joseph Weber, a physicist at the University of Maryland, resurrected Einstein's prediction and began searching for possible signals caused by GWs. He thought he saw evidence for them in meter-long resonating metal cylinders,[14] which launched a minor industry around the world as many groups attempted to repeat his results. But by the mid-1970s it was becoming clear that resonant metal bars would not be sensitive enough to detect GWs, and the GW community was primed to try something else.

The seed of LIGO was planted in the summer of 1975 when all the hotel rooms in Washington, DC were booked by tourists on the same day the general relativity theorist Kip Thorne arrived from Caltech for a meeting. Rainer Weiss, an experimental cosmologist from MIT, picked him up at the airport, so they ended up sharing Weiss' room, staying up all night talking about the future of general relativity research.[15] Thorne was a leading theorist in the physics of GWs, and Weiss had been tinkering with interferometers since the late 1960s. He had written a report in 1972 to NASA that outlined the essential challenges of using an interferometer to detect GWs.[16] That night Thorn and Weiss decided on a bold plan for MIT and Caltech to build large ultra-stable interferometers to detect the waves as they passed through the Earth. From Thorne's understanding of the exceedingly small strains caused by the waves, it was clear that the interferometers would need to perform at levels far beyond the reach of current technology, but not so far as to preclude eventual detection.

That same summer, Peter Kafka of the Max Planck Institute in Munich Germany received a request by NSF to review a grant proposal from Weiss proposing to develop interferometers that may one day be capable of detecting GWs. When Kafka later gave an overview talk on the field of GW detection, including the idea of interferometry, the entire thinking about GW detection underwent a sudden phase transition, nucleated by the idea of interferometric detection, which swept through the GW community that was already looking for the next big idea after the Weber bars had failed. Soon there were experimentalists in Germany and Italy pursuing the idea, including an especially creative experimentalist in Scotland.

Ronald Drever (1931–2017) was educated at the University of Glasgow where he became a faculty member. In 1975 he was in the audience when Kafka gave his overview, and within three years Drever had developed a large, sensitive Michelson interferometer with 10-m arms in Glasgow. Meanwhile, Weiss had started his own modest development of a 1.5-m interferometer, and he continued to discuss the project with Thorne, alerting Thorne to the progress that Drever was making in Glasgow. In 1978, with Weiss' tacit approval, Thorne invited Drever to move part time to Caltech. Within a year, Drever had a large interferometer up and running.

Weiss approached Richard Isaacson, the program manager for gravitational physics at NSF, to propose a feasibility study of a large kilometer-scale interferometer for GW detection with the possibility that NSF might fund the facility using a funding model similar to large particle accelerators. Isaacson was receptive, and Weiss with coworkers spent the next three years developing the study. When it was ready in 1983, it outlined the basic details of the construction and operation of a large interferometer. This groundbreaking study, fondly referred to by those in the business as "the Blue Book,"[17] turned the wild talk during that sleepless night with Thorne in Washington into a real plan ready for initial funding. One caveat imposed by NSF was that the MIT group and the Caltech group would need to merge, because NSF would only fund a single project at this scale. The Caltech–MIT project launched in 1984 with seed funding and was named the "Laser-Interferometric Gravitational Wave Observatory," or LIGO for short. But it had no sooner launched when it hit its first shoal—competing ideas between Weiss and Drever.

An early sticking point in the interferometer design was the technical question of how to increase the effective arm length of the interferometer, which was needed if they were to have any hope of success. The strain, h, of a "typical" GW would be astronomically small in the range of 10^{-21} at a frequency around 100 Hz, and the change in length of one arm of the interferometer, L, would be proportional to the length through $\Delta L = hL$. The thinking at that time was that for optimal detection L would be 750 km, or about a quarter of a wavelength of the GW, displacing the interferometer mirrors by an astoundingly small 10^{-18} m, or about a thousandth of the radius of a proton. Just as Michelson had developed his multi-pass configuration to produce larger arm lengths, bouncing light beams back and forth between multiple mirrors in the Michelson–Morley experiment, Weiss proposed using a mirror cavity with a small entrance hole in one of the mirrors in a configuration known as a Herriott delay line. The delay line had the important advantage that it worked for any wavelength, making it broadband and easy to reconfigure if the light source were later changed.

Drever, on the other hand, proposed using a Fabry–Perot configuration, letting the light waves build through superposition to large power and long effective distances. The Fabry–Perot size was considerably smaller than the delay line, and it was less prone to light scattering. It also was compatible with another key invention that Drever had made called power recycling, as well as with several advances that came out of the German GW group in Garching, outside Munich, called mode cleaning and signal recycling. On the other hand, the German group had already demonstrated that Weiss' delay line could work. The uncertainty about how to proceed created an impasse between Weiss and Drever, on top of other technical disagreements, which slowed progress.

In view of these uncertainties between Drever and Weiss, NSF insisted that a professional program director be put into place to guide such a large project. It was not just the problem of resolving internal conflicts, it was viewed that the troika of Thorne, Weiss, and Drever was not equipped to handle the scale of the logistics of the large facilities. At this time Rochus (known as Robbie) Vogt stepped down as provost of Caltech and became available to direct LIGO. Vogt brought a new level of professionalism to the organization, but unbeknown to everyone at the start, he also brought a tendency to micro-manage. One of Vogt's early important decisions was to adopt Drever's Fabry–Perot approach as the official design of the interferometer. But this finding by Vogt in favor of Drever was no harbinger of easier times, because there was a fundamental, cataclysmic incompatibility between Vogt and Drever that put them on a collision course that would wreck them both.

Drever was short in stature, with the face of a bulldog and the nature of one too, tenaciously grabbing onto an idea and never letting go. He also had the hint of the idiot savant, possibly being at the highly able limit of the autism spectrum.[18] His communication skills were awkward and slow, while his mind raced ahead with endless new ideas, half of them crazy and half of them genius. He had an idiosyncratic way of working that looked like chaos, but which moved fast and got results. He worked by intuition rather than by strict design, although he was very good at designing on the fly too, making choices (usually the right ones) even when there was no direct way of knowing what would work. It was this talent that allowed him to succeed in Glasgow so quickly with the 10-m interferometer, and why he was invited to Caltech by Thorne to work on a 40-m interferometer—Drever had it working within a year. Unfortunately, this haphazard approach of trial and error, but with fast turnaround, was completely anathema to Vogt's top-down management style.

One of the changes that Vogt brought into the project was weekly progress meetings where everyone reported the accomplishments of the previous week. When Drever presented the work of his team at these progress meetings, Vogt became increasingly annoyed as Drever tinkered and tinkered and tinkered, week after week. As time went on, Vogt became more belligerent and started shouting at Drever in front of the group out of frustration. Large science cannot survive endless tweaking, so Vogt began to marginalize Drever, removing him from the critical development lines of the larger project. Drever never understood this. He struggled to keep contributing to LIGO but was pushed farther out of the way until it all finally came to a head.

In 1992, one of Drever's colleagues in Glasgow, Brian Meers, was extending Drever's earlier idea of power recycling. Power recycling is a cavity-within-a-cavity approach that reflects the power coming out of the symmetric port of the interferometer back into the interferometer, making it possible to use low-power lasers to achieve high powers inside the interferometer—essentially creating a Fabry–Perot that has another Fabry–Perot as one of its end mirrors. Drever and Meers had been discussing some of the details on and off when tragedy struck. Meers and a friend were climbing Scotland's highest peak, Ben Nevis, in foul weather when they fell to their deaths from a ledge. A few months later a special session was organized for an upcoming GW conference in honor of Meers to which Drever submitted an abstract on the topic of recycling.

Around this time, Vogt set a policy that all talks by LIGO members must go through him. He would either assign someone to give a talk, or he would give the talk himself as sole spokesperson of the collaboration. He took an even harsher stance toward Drever, telling him that he could no longer represent LIGO in any meeting, even if he was invited to talk. Drever argued that his talk on recycling was independent of LIGO—it was about his Glasgow colleague after all—and the technique had not been adopted for the official LIGO design (at that time). On his return, he was called into Vogt's office and was fired from the LIGO collaboration, asked never to set foot again on the premises except to retrieve his personal belongings from his office, and then only with an escort. At the same time an email was sent to everyone in the collaboration that Drever was not to be allowed access to any LIGO facilities or resources.

Everyone but Drever saw this coming.[19] As ingenious as Drever was, he was difficult to work with, was pigheaded, and was a constant disruptive factor in meetings. On top of that, he had driven Vogt nearly to hysteria. By the end, Vogt could barely stand being in the same room as Drever, walking out of meetings if Drever entered the room. As powerless as Drever was to defend himself, he had strong supporters who pushed back against Vogt as well as lobbying the administration at Caltech to reinstate him. The foul situation led to meetings and committees and hearings and arbitration, Vogt pushing back as he was pushed further into a corner—until Weiss had had enough. By 1994 the program managers had lost patience with the dysfunction and the lack of progress. Weiss told Vogt that he had to go.[20]

Around this time the Superconducting Supercollider (SSC), which would have detected the Higgs boson years earlier than it later was, had just been canceled by Congress, leaving a large number of US particle physicists without a project to work on. One of these was the Caltech physicist Barry Barish, who was asked to take over as project director of LIGO. NSF was about to pull the plug on the entire project, but somehow Barish magically convinced them not only stay the execution, but to sink far greater amounts of money into it—up to $350 million for a facility that would not even be sensitive enough to detect GWs. As crazy as this sounds, it was part of the plan that Barish brought to the project that made it fundable. Although the initial LIGO would not be sensitive enough to detect the elusive waves, it would be built with future upgrades in mind that could be added bit by bit without changing the facilities, until an advanced form of LIGO could be established that finally could do the job. NSF agreed and released the funds to begin construction.

The first light for the LIGO facility came in 2001 as the two sites, one in Livingston, Louisiana, and the other at Hanford, Washington, began engineering runs followed by observational runs starting a year later. The performance sensitivities were dismal—far too rough to be able to detect any expected GW source—but that was exactly what was expected and what had been planned for. The quarry that LIGO sought to catch was so elusive, so incredibly faint, that measurements would approach the sensitivity floor by successive iterations, improvement by improvement, making observational runs interspersed with engineering upgrades. By 2009, after a decade of steady progress, the sensitivity floor had been pushed down by an incredible factor of 100,000 relative to the floor at first light. This accomplishment is a testament to the talent of the LIGO scientists and engineers. But no GW sources were detected, and Initial LIGO had reached the end of what that phase of the design could achieve.

In November 2009, I invited David Reitze, the spokesperson of the LIGO collaboration at that time, to give the physics collo-quium at Purdue University. Reitze was well known in the optics community because of his background in ultrafast laser physics. He had been appointed as the scientific spokesperson of the LIGO collaboration in 2007 and became the executive director in 2011.

In his colloquium, he described the design principles of the LIGO interferometers and gave an overview of the types of gravitational sources that could be detected. I have seen many physics colloquia over the years, and the character of Reitze's colloquium up to that point followed a standard template that was all too familiar from large collaborations, such as in high-energy particle physics, as they work on their next-generation detectors before they have results. Almost any talk can be swapped for any other, often with exactly the same slides and a lot of what is called "boilerplate." But then Reitze introduced us to Advanced LIGO.

Advanced LIGO would be a substantial technology upgrade that was to start the following year in 2010. Rather than a small increment, it was a major overhaul focusing on the stability of the interferometer. The details were mostly mundane, but the prospects were astounding! Where Initial LIGO had slowly pushed the noise floor down by a factor of 100,000 from its starting point in 2001, Advanced LIGO was only going to push it down by an additional factor of 10. As modest as that seemed, Reitze then showed a three-dimensional map of our corner of the universe. On the map was a small sphere showing the distance out to which Initial LIGO could detect events. Surrounding that small sphere was a large sphere that was the new limit for Advanced LIGO. The point that Reitze's slide drove home was the trivial yet profound point that a factor of 10 in radius is a factor of 1,000 in volume that can be probed. On the same slide was a simple table stating the expected rate of gravitational events per volume. Given a rate of events that would take Initial LIGO a thousand years to make a single detection, Advanced LIGO would detect one event per year. I have rarely been impressed by talks that have no results, but I was thoroughly impressed by that one slide. At the end of the colloquium, after Reitze had been mobbed by graduate students asking questions, we left the colloquium lecture hall together and I knew that I had glimpsed a future Nobel Prize.

The final design of the Advanced LIGO interferometer, that combines all of the optical tricks needed to tease out a strain of 10^{-23}, is called a dual-recycled Fabry–Perot Michelson interferometer, shown in Figure. 6.5, that incorporates Drever's

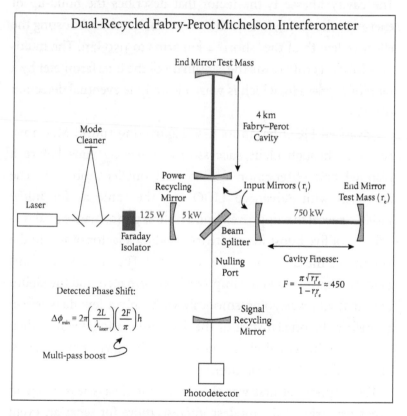

Figure 6.5 Advanced LIGO dual-recycled Fabry–Perot Michelson interferometer. A laser power of 125 W builds to 750 kW in the 4-km-long Fabry–Perot cavity.

contributions of power recycling and Fabry–Perot cavities. The laser is a Neodymium-doped Yttrium-Aluminum-Garnet (YAG) laser operating at a wavelength of 1.06 microns in the near infrared. The mode cleaner, pioneered by the German GW observatory called GEO600 outside Hanover, is a stabilization technique. The input power into the power-recycling mirror is 125 W that is boosted by the first-stage cavity to 5 kW and is boosted again to an impressive 750 kW in the main Fabry–Perot cavity. The cavity finesse F, the factor that describes the build-up of energy in the cavity, has a value of several hundred, boosting the effective length of the "short" 4-km arms to 1150 km. The multi-pass Fabry–Perot boosts the sensitivity of the interferometer by a ratio of $2F/\pi = 286$, which is what allowed the eventual detection of GWs.

Advanced LIGO began its first engineering run in May 2015, working through eight successive engineering runs before a planned first observation run to start on September 18. The approach with Advanced LIGO was the same as for Initial LIGO—successive improvement. The first performance goal was a factor of five boost in sensitivity, yielding a factor of 100 in the volume of the universe to be sampled.[21] The engineering runs quickly demonstrated the improved performance, and the eighth and final run was going smoothly with only a few days left to go before the official start of the first observational run[22] when Drago saw his email alert and sent out his own, now famous, email to the full collaboration.

The response at first was the same as it always was when an event was triggered—modest interest, more for what an event might say about the performance of the interferometer than any belief that the event might be real. The alarm rate at LIGO was purposely set to give false alarms at the rate of several per month. Too many false alarms and you run into the "crying wolf" problem; too few and you might miss something real. If anything, the alarm rate was set a bit too close to the "crying wolf" side, so there was a general ennui about new events.[23] In addition, it was standard practice to sometimes inject a synthetic signal into the detectors to test the detection and analysis algorithms, yet the synthetic signal injection equipment had been turned off for the engineering run and was not a possibility. Furthermore, for this event there was a strong correspondence in the data from the two LIGO sites located in Livingston, Louisiana, and Hanford, Washington. Despite the noise in the strain data, as there must be when searching for signals at the limit of detection, there was a definite form to the pair of signals that looked just like they were supposed to. Excitement grew over the next two days, and then one of the analysis groups requested that the event officially be considered a GW candidate.

The official request, and its acceptance by the LIGO management, launched a rigorous process that locked down the performance of the detectors—no changes of any kind could be made to any of the hardware or software systems. Then began a month-long background run were the detector measured nothing but noise, rigorously quantifying the noise floor under exactly the same conditions as for "the Event" while looking for spurious signals that might masquerade as real. A small task group even explored the miniscule but non-zero chance that the LIGO computer system had been hacked and that the signal was a prank, but the task group was able to rule that out. By the end of the month the LIGO teams had quantified the false alarm rate for such an event to one false detection in 22,500 years. This also placed an uncertainty of 4.5 sigma on the event. The computational models by this time had pegged the event as the merger of two medium-size black holes, each of about 30 solar masses, which had been captured in the last seconds of their in-spiraling orbit. Just prior to their coalescence, each was traveling relativistically at roughly half of the speed of light. In that second just before the merger, the system radiated away the energy equivalent of three solar masses, pumped it into the warpage of space-time, as the GW started its billion-year journey outward where a tiny fraction of the wave eventually would be intercepted by the Earth.

An uncertainty of 4.5 sigma is often good enough for a scientific discovery, but over the years there have been cases when even a 6-sigma statistic eventually disappeared under more data—to the great embarrassment of the researchers who initially announced such a discovery. Therefore, everyone who was part of LIGO was sworn to complete secrecy. The political cost of getting this wrong would be too great. Confidence increased on October 12 when another very weak GW signal was detected. This was just after the lockdown had finished, and the observational run named O1 was underway. Then, the day after Christmas, another strong event was detected, this time from 14- and 8-solar-mass black holes also about a billion light-years away. This third event cemented confidence in the first event, and the LIGO collaboration called a press conference for February 11, 2016 to be held at the offices of the NSF where it was streamed live. I remember hearing about it in the hallway after coming out of class, and about 30 of us crammed into a small room in the physics building at Purdue to watch the news conference live. David Reitze, by then the

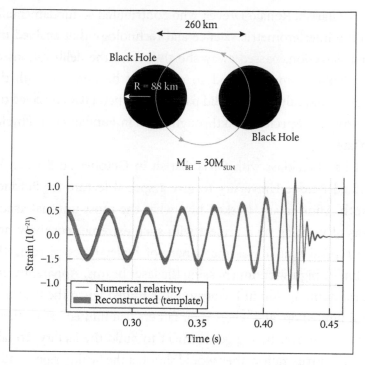

Figure 6.6 First detection of a black-hole merger using the LIGO interferometer systems. As the two black holes spiral inward, they radiate gravitational waves, and the frequency increases as the two bodies approach each other and merge. The image at the top shows the relative distances as the signal begins.

https://www.ligo.org/science/Publication-GW150914/images/fig-3.png

executive director of LIGO, made the announcement, remarking "It's the first time the universe has spoken to us in gravitational waves,"[24] as he unveiled the signal (reproduced in Figure. 6.6). I had been anticipating this day for six years ever since I had walked with Reitze down the hall after his colloquium.[25]

It took little more than a year for the Nobel committee in Sweden to award the prize to Ranier Weiss, Kip Thorne, and

Barry Barish. Ronald Drever, who contributed so fundamentally to the interferometric science and technology that enabled the first detection, passed away shortly before the deliberations of the Nobel committee and so could not be honored with the Nobel Prize, although he did participate when LIGO received the prestigious Special Breakthrough Prize in Fundamental Physics in late 2016.

I had breakfast with Barry Barish in October 2018 when he visited Purdue University to give a special lecture for Purdue's 150th Anniversary. I asked him what the most unusual aspect was of working on a project with such extreme sensitivity, and he replied that it was the danger of bullet holes piercing the 8 km of high-vacuum encasing the laser beams. Apparently, the local hunting club in Livingston was miffed when the land they had hunted on, and where their fathers had hunted before them, was taken over by big government to build the facility. To take out their frustration, they would shoot at the facility signs, and at least one stray bullet had hit the building. Barish, as the director of LIGO, asked for advice how to settle the problem and was given two suggestions: bring the problem to the FBI, or talk to the local sheriff. The project management recommended going to the FBI, because it was a national facility. When he tentatively approached the FBI, they outlined a massive response to build a bunker surrounding the site topped with high fences, so Barish decided instead to phone the local sheriff. The sheriff asked Barish to meet him at a local dive, and when he arrived, he found a full table—the sheriff had invited the hunting club. Barish has a characteristically cool manner. He is tall and charismatic with a wise man's white beard surrounding an easy smile, carrying himself like a paternalistic grandfather putting you at ease. He bought the guys a few rounds of beer, made friends, and invited them to tour the facility. After that the bullets stopped, and the hunting club members became vocal local supporters of LIGO. This story gives insight into Barish's deft ability to solve logistic problems together with his sensitivity to people. Barish left LIGO temporarily in 2005 to lead a next-generation particle accelerator project, leaving behind a professional organization and facility that was making steady progress, and he returned in 2012 as they began their sprint toward that first event a few years later.

Since "the Event," there have by now (mid-2022) been over 20 gravitational events detected, as well as an equal number of marginal detections with low signal-to-noise, many of which may also be legitimate events.[26] The confirmed detections include the largest black hole pairs (66 and 85 solar masses) and the smallest (8 and 11 solar masses). Recent events also include neutron binary mergers (1.3 and 1.5 solar masses) and neutron-star and black-hole mergers (2 and 9 solar masses). The majority (77 percent) of confirmed detections are binary black-hole mergers. Binary neutron-star mergers (9 percent) and neutron stars merging with black holes (9 percent) make up most of the rest. One event in 2019 involved a black hole (23 solar masses) with a companion whose mass (2.6 solar masses) make it ambiguous whether it was a neutron star or a very small black hole.

The European VIRGO GW detector in Italy has participated in the detection of many of the recent events, and the LIGO-India interferometer will soon be up and running (in 2023 or 2024), and the KAGRA interferometer in Japan began running in 2020, providing a worldwide network that can locate the direction of the signal to sufficient accuracy that conventional telescopes can be trained in the same direction to look for electromagnetic evidence of the same event. Merging gravitational and electromagnetic channels is called multi-messenger astronomy, which is a new form of astronomy that is just beginning. For instance, GWs from the double neutron-star merger designated GW170817 was detected by both LIGO sites as well as VIRGO on August 17, 2017. The three arrival times allowed the general direction of the event in the sky to be triangulated. At nearly the same time that the GW was recorded, gamma rays were detected by astronomical satellites from the same region of the sky. Then, 11 hours later, multiple telescopes observing across the electromagnetic spectrum using radio waves and infrared as well as ultraviolet and x-rays detected a bright transient event in a galaxy 140 million light-years away in the constellation Hydra. All the observations were consistent with the theory of neutron-star mergers, and this was hailed as an inaugural GW event for multi-messenger astronomy, being selected as the breakthrough of that year by *Science* magazine.

So far, no merger involving a white dwarf has been seen because these events are weaker and at lower frequencies, which makes it difficult to detect them with the current Earth-based interferometers. For this reason, ESA has taken initial steps to put an ultra-large-baseline interferometer into space. The interferometer, known as LISA (Laser Interferometric Space Antenna), will consist of three satellites flying in formation at the L1 Lagrange point of the Earth–sun system on a triangular baseline of 2.5 million km. A proof-of-principle mission, known as LISA Pathfinder, was launched on December 3, 2015 (only three months after "the Event" and before it was even announced) from Kourou, French Guiana. The spacecraft operated for a year at the L1 Lagrange point, achieving a noise performance better than the design value. Based on the success of this test flight, the full LISA mission is planned for launch in 2034. LISA is expected to detect white dwarfs being eaten up by black holes, binary white-dwarf mergers (there are thousands of binary white-dwarf systems in our own galaxy), black-hole binaries that will be detected orbiting each other years before they merge, continuous-wave sources from spinning neutron stars, and even the GW echo of the Big Bang—future discoveries all relying on the power and the finesse of optical interferometry.

Two Faces of Microscopy

Diffraction and Interference

The generated (microscope) images are the interference phenomena which accompany the diffraction effect of such objects.[1]

Ernst Abbe, 1878

D iffraction and interference are so closely coupled that it
is sometimes hard to separate them conceptually. All
diffraction relies on interference, but not all interference relies on
diffraction, making diffraction a subset of interference phenomena. The first interferometric system, Young's double slit, relied
intrinsically on diffraction, and Fresnel's study of diffraction led
him to confirm the wave nature of light. Fifty years after Fresnel, the German physicist Ernst Abbe, working with Carl Zeiss,
the German microscope manufacturer, developed a theory of
imaging based on light diffraction that set the ultimate limit for
microscopic resolving power using light. Abbe's diffraction limit
of microscopic imaging stood for a hundred years, seemingly an
unsurmountable barrier, until it was finally surpassed by enlisting
even deeper physics of diffraction and interference that originate
from simple periodicities pioneered by an unschooled German
autodidact.

Interference. David D. Nolte, Oxford University Press. © David D. Nolte (2023).
DOI: 10.1093/oso/9780192869760.003.0007

The Rules of Diffraction

In the summer of 1801, a glass-cutter's workshop and house collapsed in the city of Munich, Germany, seriously injuring the glass-cutter's wife and pinning a 14-year-old orphan apprentice in the rubble. The sensational collapse caught the imagination of the public, and it even reached the attention of the future Elector of Bavaria, Maximillan Joseph, who visited the site to watch the rescue work. The boy was extracted after many hours, miraculously unscathed. Moved by what he saw, the prince-Elector asked his privy counselor, Joseph von Utzschneider, to provide financial assistance to the boy. This chance meeting between a knight of Bavaria and a hapless boy would later propel German glassmaking to the pinnacle of international trade in high-precision optical instruments. The boy was Joseph Fraunhofer.

Joseph Fraunhofer (1787–1826) was born in a small town in Bavaria to a poor glazier and his wife. When Fraunhofer was 11 years old, his parents passed away and his guardians apprenticed him off to the glass-cutter in Munich. The glass-cutter was a surly and uneducated man who did all he could to dampen Fraunhofer's attempts to educate himself, keeping him from attending school and even stopping him from reading at night. The naturally intelligent and curious Fraunhofer felt stifled, with no foreseeable way out of his suffocating situation. Then, when the ill-kept workshop collapsed onto him on July 21, 1801, actually suffocating him, it was as if a prayer had been answered. After the boy was pulled from the rubble, von Utzschneider granted Fraunhofer the sum of 18 ducats with which Fraunhofer bought books on optics, purchased an optical grinding machine, and paid for release from the remainder of his apprenticeship.

Five years later, Utzschneider needed a skilled optician to help him set up an optical manufacturing operation in Benediktbeuren,[2] a former monastery located some 50 miles south of Munich in the rolling foothills of Bavaria. The property had become available for purchase from the Bavarian government after the forced nationalization of all monasteries within the kingdom. Fraunhofer's fortuitous acquisition, with Utzschneider's help, of the glass-grinding machine and his books on optics, to which he had applied himself industriously, now made him the ideal candidate for Utzschneider's new business venture. This situation could not have worked out better even if it had been planned. Fraunhofer began working at Benediktbeuern in 1806 making high-quality glass for the Reichenbach Mathematical-Mechanical Institute in Munich, and in 1809 he was put under the supervision of a Swiss glassmaker, Pierre Louis Guinand.[3]

Fraunhofer and Guinand worked successfully together but with some tension. Guinand was a craftsman who labored intuitively, while Fraunhofer was painstakingly systematic, performing carefully planned experiments to steadily improve the quality of the glass and lenses they were making. The two conflicting methods put them at odds with each another. The division of labor left the glass melting entirely to Guinand, while Fraunhofer was responsible for all the downstream processes in the optical cutting, grinding, and polishing. Unfortunately for Guinand, the increasing precision of Fraunhofer's lens-making exposed deficiencies in the glass melting which Guinand was slow to improve. Therefore, by 1811, Utzschneider had reorganized the Benediktbeuern operation, pushing Guinand aside and putting Fraunhofer in charge of the entire manufacturing process, making him a named partner in the venture—henceforth, the Utzschneider–Franhofer Optical Institute.

Now free to operate as he wished, Fraunhofer pursued two coupled lines of research—one fundamental and the other applied. The fundamental research was on the physical properties of light interacting with glass, while the applied research was on the steady improvement in the quality of glass manufacturing. These two lines were closely linked, as better understanding of one helped the other and vice versa. It was during this time that Fraunhofer made his famous—and useful—discovery of dark lines in the spectrum of sunlight. These are caused by the absorption of the thermal light of the sun by elements in the outer solar atmosphere and are known today as Fraunhofer absorption lines. The dark lines had been noted by the British chemist Wollaston a decade earlier, but Fraunhofer studied them systematically and with high precision. He also used them as wavelength standards in his studies of the refractive indices of different types of glass.

One of the outstanding problems in high-precision optics in Fraunhofer's day was the construction of achromatic lenses.[4] A single lens always produces a rainbow-like focal spot in white light because chromatic aberration causes different colors to refract differently in the glass lens material, making images blurred and tinted. Many years before, around 1730, Chester Moore Hall, a British lawyer and inventor, conceived of a way to form a double-lens from two different types of glass, known as crown glass and flint glass, one with a lower refractivity and the other with a higher, to compensate the chromatic aberration, as shown in Figure. 7.1. Such achromatic lens doublets were difficult to make because slight differences from one batch of glass to another would ruin the compensation. Fraunhofer, with his superior

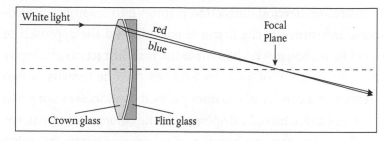

Figure 7.1 Achromatic doublet lens. Different colors are refracted differently by the different types of glass to form a single focal image for both red and blue light.

lens-making abilities and control of the glass-melting process, plus having wavelength standards (the Fraunhofer lines), was in a perfect position to systematize the fabrication of high-quality achromatic doublets of increasingly larger diameters. These achromats became the centerpieces of a bourgeoning business for Benediktbeuren in astronomical telescopes.

Fraunhofer's systematic study of the science and practice of making achromatic lenses led to his first publication in 1817 (where he also reported on the dark lines of sunlight and their uses in precision refractometry).[5] His advances brought this self-educated scientist to the attention of the scientific community, and he became a member of the Bavarian Academy of Sciences in 1821. In 1824 he was named a peer of Bavaria—Joseph Ritter von Fraunhofer. During the time when he was receiving these accolades, he was extending his work on refractometry by creating spectroscopic devices with much greater dispersion than glass—diffraction gratings.

When Fraunhofer began experimenting with diffraction gratings, Fresnel's publications had just become available in Germany, allowing Fraunhofer to use the new wave-based principles to calculate and develop diffraction gratings into scientific spectroscopic instruments. His first attempt followed the approach of David Rittenhouse who had made the first diffraction grating by stringing an array of 50 hairs between the fine threads of two screws with a density of 100 lines per inch. Fraunhofer strung 260 fine wires and achieved a dispersion of the colors of light that was already greater than his best flint-glass prism. Despite the utility of such a diffraction grating, stringing fine wires was a difficult and tedious process which quickly reached a limit for the number of wires that could be strung practically. Fraunhofer began to look for a way to produce diffraction gratings more easily and with greater dispersion.

By this time, Fraunhofer was arguably the finest optical instrument maker in continental Europe with the resources of an entire optical institute at his disposal, including fine mechanical engraving tools. He modified a diamond-tipped inscriber to produce finely ruled lines, first on glass, then on glass covered in gold foil, and finally onto smoothly polished metal, increasing the brightness and the dispersion of the gratings. The line density of the gratings reached 3,200 lines per Paris inch, representing a spacing of only 8.4 micrometers with the dispersion to deflect green light by 60 mm across a distance of 1 m. In comparison, Rittenhouse's grating deflected light by only 2 mm. Fraunhofer published the work in 1823 in *Annalen der Physik*.[6]

Though he continued working to implement grand plans for the glass works, Fraunhofer's seemingly infinite energy was finally sapped by overwork and too many hours breathing the poisonous lead fumes of the flint glass ovens. His already frail health failed him in 1826, and he died at the early age of 39 consumed by tuberculosis. Within a few years, the dominance in the fabrication of precise diffraction gratings passed from Germany to the nascent technical industry in the United States where gratings achieved an apex in the late 1800s under the skillful hands of the first president of the American Physical Society.

Henry Augustus Rowland (1848–1901) was born in Pennsylvania, coming from a long line of Presbyterian ministers, but he broke the chain to become a scientist. When he attended college at the Rensselaer Polytechnic Institute, he found the other students profane (given his family business), but he excelled at his studies and returned to Rensselaer a few years later as an instructor of physics. In his free time, he performed original research on the physics of magnetism that brought him praise from Maxwell in England and a job offer to be the first professor of physics at Johns Hopkins University, founded in 1876 modeled on the European research universities. Rowland's first task was to travel among the universities in Europe to observe how they worked and to make a list of research equipment to purchase for the university's new state-of-the-art laboratories.[7] The premier physicist in Europe at that time was Helmholtz, and Rowland spent several months in Helmholtz' lab in Berlin only four years before Michelson's memorable visit in 1880. Those months were surprisingly fruitful. With Helmholtz' equipment, Rowland performed an experiment he had devised several years earlier to test whether a moving static charge could produce a magnetic field as predicted by Maxwell's equations. The experiment was a success, and Rowland's reputation in America as well as in Europe was established.

When Rowland returned to Johns Hopkins, he set up his labs and took on a series of physics students who obtained their doctoral degrees from him. One of those students was Edwin

Hall who performed an experiment to test whether moving charge in conductors was deflected by static magnetic fields—a phenomenon known today as the Hall effect. Many of Rowland's students went on to populate the first chairs of physics in university departments across the states, which is why Rowland is often called the father of American physics. He was also a founder of the American Physical Society (APS) in 1899 and was elected its first president, with Michelson as its first vice-president.[8]

Well into his first decade at Hopkins, Rowland was exploring magnetism and conductivity as well as thermodynamics when his attention was diverted by a simple piece of lab equipment that he had purchased for an astronomy colleague—a diffraction grating. This particular grating had been fabricated by the amateur astronomer Lewis Rutherford using a ruling engine driven by a high-precision screw drive. Rutherford had been making the most precise diffraction gratings since 1871, and they were used by astronomers around the world in the young field of astronomical spectroscopy. One day, while Rowland was riding the train to Washington and talking with a colleague from Hopkins, he had a sudden insight into how to rule lines onto a curved surface of metal.[9] Such a curved diffraction grating would act as its own focusing element and eliminate the glass lenses that were used in all spectrometers of the day. Optical absorption in the glass distorted the measured spectrum and decreased the intensity, lengthening measurement times. The curved surface also made it possible to measure much broader bandwidths, further decreasing measurement times. In addition, Rowland came up with an idea to improve the precision of Rutherford's screw drive.

With one of the new curved gratings in his suitcase, Rowland attended a conference in Europe in 1882 where he unveiled his new invention. The curved grating had a radius of curvature of 21 feet and was ruled with 29,000 lines per inch, able to deflect green light by 600 mm over a distance of a meter compared with Fraunhofer's 60 mm. With the new grating, Rowland acquired the most detailed measurement of the solar spectrum ever made, completing the measurements in only a few hours compared with several days when using the older gratings. The gratings created a sensation at the conference, and there were immediate requests for the new instrument from observatories around the world. Rowland entrusted the work on the new ruling engine to Hopkins' highly capable instrument maker Theodore Schneider who made hundreds of these gratings, receiving a royalty on the proceeds for his efforts, while Rowland sold them all at cost, making no money for himself or for Hopkins.[10] In the following decades, several hundred gratings were distributed to observatories across the globe, and they remained the dominant spectroscopic instruments well into the twentieth century.

Rowland's gratings were the technological legacies of Joseph Fraunhofer's first ruled diffraction grating a century before. Fraunhofer's other legacy was the dominance of German glass manufacturing and high-precision optics that extends to this day. Even though by mid-century the Fraunhofer glass works had languished without their indefatigable founder, and the workshop had been dismantled, a new generation of glass and optical craftsmen emerged in a city 400 km due north from Munich.

German Glass

The German city of Jena is famous today for its nearly 200-year history of optics manufacturing. It resides on the western edge of the state of Thuringia, a three-hour drive south of Berlin on the autobahn that speeds you through rolling countryside of tree-covered hills punctuated by neat cultivated fields. On the outskirts of town, a trail winds up the side of a wooded hill to a marker commemorating the vanguard skirmishes of the battle of Jena in 1806 when Napoleon's army swept into Prussia. The battle was lost by troops under the command of the Duke of Brunswick, who happened to be the patron of the mathematician Friedrich Gauss. The Duke was mortally wounded in the battle, forcing Gauss to find employment at the University of Göttingen where he influenced a generation of German mathematicians that included Bernhard Riemann. At the base of the hill, the Carl Zeiss Microscopy GmbH research office is a crisp silver and glass building that looks like a modern monument to light. Yet the true monument is in town, at the Deutsches Optisches Museum, where the works and artifacts of Carl Zeiss (1816–1888) figure prominently in the museum. Sharing equal honor of place in the museum are highlights of his right-hand man.

Ernst Abbe (1840–1905) is the physicist most associated with the theory of microscope imaging, and his name is attached to the equation for the resolution limit of microscopes—the size of the smallest feature that can be resolved using ideal optics. Abbe was born into a family of little means from a small town outside Jena. Fortunately, his father had a steady job at the local spinning mill, and his father's employer sponsored Abbe with a scholarship to attend the college preparatory gymnasium.

Because of Abbe's talent and determination, his father decided to pay his tuition to the local University of Jena and later at the University of Göttingen, where Abbe took courses under Riemann and Weber and received a doctoral degree in physics in 1861. He returned to the University of Jena, first as a lecturer and then as an assistant professor of mathematics, physics, and astronomy.

During the early 1860s, the scientific instrument company of Carl Zeiss had begun successfully selling high-quality compound microscopes to buyers across Europe, but each unit was custom crafted by fitting together lenses in a trial-and-error manner that was time consuming and expensive. Zeiss had engaged a series of consultants to try to systematize the process so that it could be converted to parallel production, but they had not been able to solve the problem. In 1866 Zeiss learned of Ernst Abbe, the new professor of physics at the University of Jena, who was known to have experience in physical optics. Zeiss hired Abbe as a consultant, and it immediately paid off.

In an excellent example of theory informing practice and of academic research supporting commercial product development, Abbe's collaboration with Zeiss produced the finest microscopes of their day. Although Abbe applied what might have seemed like esoteric principles of theoretical physics, he systematically derived for the first time how light propagates through optical imaging systems, and his findings were quickly put into practice by the company in the manufacture of improved optical elements. For instance, Abbe discovered what is known as the sine law of optical focusing which is the condition on the curvature of a large lens that must be met to form a minimum focal spot size. This discovery allowed the Zeiss craftsmen to grind lenses with reduced spherical aberration, and it also achieved the long-standing goal of systematizing the design of lens systems in compound microscopes by replacing the trial-and-error approach.

Abbe improved on Fraunhofer's achromatic doublet lens by creating an apochromatic triplet lens that removed higher-order chromatic aberration. This advance further improved the performance of the Zeiss microscopes and raised the stature of the company, becoming the top microscope manufacturer in the world. Because of his successes, and their close association, Zeiss made Abbe a shareholding partner in 1876, and after Zeiss passed away in 1888, Abbe became the owner and director of the company. In his final career move, Abbe left his post at the university in 1891 to launch the Carl Zeiss Foundation, which was an early example of a social welfare organization that became a model for later progressive social programs across Germany.

After exiting the front door of the Optisches Museum, stepping from dark into sunlight, you can cross the Carl-Zeiss-Platz to a small city park dominated by a surprisingly impressive monument. This is the Ernst-Abbe-Denkmal, or memorial, a small building containing four bronze reliefs representing industry, mining, agriculture, and shipping. A tall white limestone pedestal in the center supports the stone bust of Abbe illuminated by a transparent light dome high above. There is a religious feeling to the place—a temple to optics built only five years after Abbe's death at a time when society held a deep respect for advanced education and had faith that science was a path to better life. Yet the most lasting memorial to Abbe—for which his name is mentioned routinely at nearly every optics conference one attends—is his theoretical resolution limit for imaging through a microscope.

A very simple question applies to any optical imaging system: What is the smallest feature of an object that can be resolved in its image given perfect lens quality? As simple as this question is, finding the answer requires a deep dive into a more fundamental question of how a lens takes the light scattered from an object to form an image. Abbe started with the diffraction of light from the object, which he envisioned as a periodic pattern of opaque lines; i.e., a diffraction grating. For light of wavelength λ incident at normal incidence on the grating with a line spacing d, the diffraction angle for the first diffraction orders is given by the same formula as for Young's double slit: $d \sin \theta = \lambda$. When the *diffracted* rays are intercepted by a lens, it *refracts* the light waves and redirects them to cross at an image plane. The crossing wavefronts form *interference* fringes whose periodic bright and dark stripes are the *image* of the diffraction grating.

To determine the smallest spacing d that a lens can resolve, Abbe imagined increasing the diffraction angle θ until the diffracted ray was at the extreme aperture of the lens. At this extreme angle, the diffracted ray is called the marginal ray of the optical system—the ray with the largest divergence angle from the object that passes through the optical system. If the diffracted angle is a little larger than for the marginal ray, then the diffracted rays cannot make it to the image plane, and there will be no interference. Therefore, Abbe concluded that the minimum spacing d that can be resolved is set by the maximum diffraction angle $d_{min} = \lambda / \sin \theta_{max}$. This simple result arising from a simple argument captures the essential physics of optical resolution. Abbe extended this simple result to a more rigorous answer by considering that the optical system is embedded in a refractive medium of index n. Furthermore, the object in a microscope is illuminated by a range of angles up to the

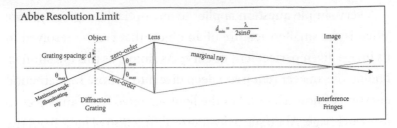

Figure 7.2 Principle of Fourier imaging by a lens. A diffraction grating with period d is illuminated obliquely at the maximum acceptance angle of the lens. The first diffraction order of the grating is at the opposite acceptance angle. The lens refracts the zero and first-order diffracted waves to intersect at the image plane where the interference forms an interference fringe pattern that is the magnified image of the diffraction grating.

maximum acceptance angle of the lens, as shown in Figure. 7.2, yielding a factor of two in the condition for the diffracted rays to cross and form interference stripes. Abbe introduced a term called numerical aperture (NA) that is the product of the refractive index with the sine of the maximum acceptance angle. Putting all of this together, Abbe's formula for the ultimate resolution of an optical microscope is

$$d_{min} = \frac{\lambda}{2n \sin \theta_{max}} = \frac{\lambda}{2NA}$$

setting the condition for the smallest feature size that can be resolved by a microscope. If the numerical aperture can approach unity for a large lens with a short focal length, this equation states that the smallest feature that can be resolved is half of the illumination wavelength.

Abbe obtained his resolution limit in 1871 when he was under consulting contract to the Carl Zeiss Werkes, so he did not openly publish the details of his theoretical findings. The influence of his work within the scientific community was disseminated mostly at informal meetings with other scientists and in oral presentations.[11] As important as his diffraction limit was for the theory of microscope imaging, his methods may have been more important. One of the most powerful theoretical approaches in optics today is known as Fourier optics. Abbe, by viewing objects as if they were a superposition of diffraction gratings, introduced the Fourier transform into physical optics. His unique viewpoint of the role of diffraction and interference in imaging inspired many successors to invent new types of microscopes, one of whom used Abbe's methods to make invisible things visible.

Invisible Things

Life is invisible, or nearly so. Place a single mammalian cell into water and it disappears, cloaked by the refractive index of water matching the refractive index of the cell. The challenge to all microscopists, from the very beginnings with Robert Hooke's 1665 book *Micrographia* and Anthonie van Leeuwenhoek's 1677 "small animals," is how to make the invisible visible. The simplest approach is to stain the sample with dyes that are taken up selectively by one cellular component or another, making the components dark and visible. This is still the dominant practice used today in the histological examinations of biological tissues. But stains are mostly toxic and almost always are applied to tissues that have already been preserved in aldehydes, preventing living cells and tissues from being viewed in real time to observe their functions.

Yet even van Leeuwenhoek, working with extremely crude optics, was able to see living single cells in water. How? The answer is that he was the first to find the solution that everyone after has found who has focused on a biological specimen. When the optical imagine is not quite perfect, either through flaws in the lenses or because the sample is not quite in focus, then the invisible object acquires slight intensity modulations around its internal and external edges. This intensity modulation is caused by unequal changes in the phase of the light as it propagates through unequal optical thicknesses in the sample. Under perfect imaging conditions, this phase is invisible in the formed image, but in imperfect conditions, the different phases of light slightly overlap and interfere, producing intensity changes, especially near sharp edges. This phenomenon was already widely known in Abbe's day, but it was viewed as an imaging artifact rather than as a path to improved contrast. Even the person who did take that path for the first time was not a microscopist but was a physicist playing with one of Rowland's famous diffraction gratings.

Frits Zernike (1888–1966) had an early life lived in sharp contrast to both Fraunhofer and Abbe. He was born into a well-educated family in Amsterdam who counted teachers among the elder members, and professors and literary figures among the siblings. Both his mother and his father taught mathematics at the gymnasium, and his father had a parallel interest in physics that Zernike absorbed as a young boy. He built his own astronomical observatory complete with a tracking mechanism for his telescope and a chemistry bench to develop his photographic plates. At the University of Amsterdam in 1908, he won a gold medal for a mathematics competition. Then in graduate school he won a competition sponsored by the Dutch Society of Sciences judged by Lorentz and van der Waals. When asked whether he would like the award as a gold medal or as cash, he took the cash, since he had already experienced the thrill of receiving a gold medal for his earlier mathematics essay.[12] This second prize related to his doctoral thesis in chemistry, awarded in 1915, on topics of statistical mechanics. Despite his chemistry training, his first university position was under an astronomy professor, and his first faculty position was in physics at the University of Groningen, where he became a full professor in 1920.

Perhaps recalling the amateur astronomical observatory of his youth, Zernike turned his research toward optics and spectroscopy once he had become a full professor and had more latitude to pursue diverse interests. In 1930 he purchased a high-quality Rowland grating that he mounted into his spectroscopy system. As part of the calibration process he used a monochromatic line source of light that diffracted at a certain angle from the grating which he observed through a sighting telescope mounted on what is called the Rowland circle. It was well known that even the most accurately machined screw in the ruling engines making the gratings—still using the process originally developed by Rutherford and Rowland—was not perfect. Therefore, gratings had a superposed long-range repeat length set by the length of the screw. The double periodicity of the ruled gratings introduced small secondary diffraction peaks adjacent to the main spectral peak. These sidelobes are called Rowland ghosts. When looking at monochromatic light diffracted from the curved gratings, the sidelobes cause a slight modulation in the intensity of the diffraction, producing the appearance of brighter and darker stripes. On the day that Zernike turned his telescope onto his new grating, he was struck by his ability to make the stripes appear stronger or weaker depending on where he focused.

Zernike quickly realized that the light diffracted in the sidelobes of the grating must have a different phase than the central large diffraction peak. As he slightly adjusted the focus of the sighting telescope, Zernike was changing the relative path lengths of the central and side peaks so that the phase offset could be made a bit bigger or a bit smaller, causing the interference to increase or decrease and the observed stripes to get stronger or weaker. From the behavior of the sidelobe interference stripes he guessed there must be a 90-degree phase shift of the sidelobes relative to the central peak. Then he had a flash of insight for how to manipulate this phase offset directly, rather than as a side effect of bad focusing. Zernike knew of a process that Lord Rayleigh had developed a few decades earlier by which a weak hydrofluoric acid solution could etch a very shallow but uniform groove into glass. He took an extremely flat plate of glass and used Rayleigh's technique to etch a groove about a millimeter wide and only a half wavelength deep. The relative refractive index of glass to air is about a factor of 2, so this "phase strip," as Zernike called it, produced a net 90-degree phase shift for the sidelobes. When he placed his phase-strip in front of the telescope lens, careful to align it perfectly on the central peak, he was thrilled to see the stripes jump out with high definition. He had converted a phase contrast into an intensity contrast.

Zernike remembered a key phrase attributed to Abbe that "the microscope image is the interference effect of a diffraction phenomenon."[13] When Abbe conceived of a sample as a diffraction grating, he was thinking purely in terms of an array of thin opaque strips. Light passing through such a slit pattern has an

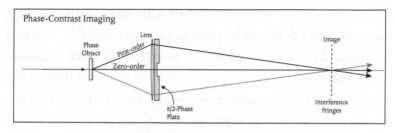

Figure 7.3 Phase-contrast microscope principle. The phase-grating object creates first diffraction orders that interfere with a zero-order at the image plane. By shifting the phase of the zero-order by $\pi/2$, the ± 1 diffraction orders cross the zero-order and produce amplitude interference fringes at the image plane, converting the phase object into an amplitude image.

amplitude modulation. Zernike, on the other hand, realized that if the diffraction grating is composed of strips of altered phase rather than altered amplitude—a phase grating rather than an amplitude grating—then Abbe's image formation based on interference would make the image uniform without modulation, just as it should. But if a phase-strip (illustrated in Figure. 7.3) were placed in the microscope, just as he had done with the Rowland grating, then the phase-grating would be converted to an amplitude grating caused by interference at the image plane, making an invisible phase-grating visible in the image. He was able to do this experiment quickly with immediately positive results. Later, when he placed transparent cells in water on the slide of his microscope, his phase-strip made them jump out in stark contrast—he had succeeded in making invisible things visible! He originally was going to call his method "phase-strip method for observing phase objects in good contrast," but that sounded too bulky, so he shortened it to "phase-contrast method."

Convinced that he had discovered something both profound and useful, and drawn by his respect for Abbe's invisible hand that had helped guide him to his new invention, Zernike contacted the Zeiss Works and traveled to Jena in 1932 to see if the Zeiss corporation would commercialize his new microscope. They would not. During the meeting, one of the old hands, one of Abbe's successors, said to him "If this had any practical value, we would ourselves have invented it long ago."[14] Zernike was nonplussed. He knew why the Zeiss people had not thought of it. Abbe—and his successors—had missed the essential point about the role of phase. Abbe had thought only in terms of amplitude gratings—repeating sections of high and low transmission. The phases of the diffraction peaks from Abbe's amplitude grating are all in phase and play no additional role in the interference effect that forms the image. Therefore, while Abbe and his followers had conceived of the initial principles of Fourier imaging, they had stopped short at amplitude and had overlooked phase.

Despite the weak reception of his phase-contrast microscope, Zernike was persistent and continued to educate the staff at Zeiss on the merits of phase contrast through the 1930s. Ironically, it was only after Zernike and The Netherlands had fallen under brutal Nazi occupation that the Zeiss company finally began manufacturing phase-contrast optics for the German war effort. The company survived that conflagration, and Zernike's phase-contrast microscopes became great workhorses in the biological and medical revolutions of the late 1940s and early 1950s. In 1953, Zernike was awarded the Nobel Prize in Physics for his invention.

Beating the Limit

Abbe's theory of microscopic imaging seemed to place a fundamental limit on the smallest feature that could be imaged in a microscope, much like the constant speed of light is a fundamental limit of the laws of physics. However, there are examples in science where supposed fundamental laws and limits were merely a mindset, and all one needed to do was think differently, and the limit could be blown past.

I remember one lecture, when I was an undergraduate student in physics at Cornell, where the professor (I confess I don't remember who) stated in no uncertain terms that single atoms on surfaces would *never* be imaged individually, because that would violate a fundamental law of physics—Heisenberg's uncertainty principle. To image a single atom, he argued, would require a stream of quantum particles with wavelengths so small that their associated momentum $p = h/\lambda$ would knock the atom off the surface.

Embarrassingly for that professor, a few years later, two young researchers at IBM in Zurich, Switzerland, dragged a needle with an ultra-sharp tip across the surface of a metal and observed individual atoms with sub-atomic resolution. Gerd Binnig and Heinrich Rohrer had just invented the scanning tunneling microscope (STM) for which they were awarded, in the same year as their invention, the 1986 Nobel Prize in Physics, and surface science has never been the same. They had broken no fundamental law of physics—it was just a new way of thinking about the problem. The Abbe limit falls into that same category—it may be the fundamental resolution limit when using a conventional microscope with conventional optical elements, but there are clever people who can think of clever new approaches, and the Abbe limit was bound to fall.

One of the groups seeking to surpass the Abbe limit was led by John Sedat at the University of California at San Francisco. As a biophysicist interested in using light to image living systems, he had been following the advances in stellar interferometry and began thinking of ways to bring those ideas into biological imaging.[15] Stellar interferometers were imaging devices (in the Fizeau configuration) that used separated telescopes, as Labeyrie had done on the Plateau de Calern, to obtain an effective aperture that is much larger than the size of an individual telescope. Sedat's idea was to use these kinds of interferometric ideas to improve the resolution of biological imaging microscopes in the far field without the need to scan point-to-point.

In 1993, a new post-doc joined his group by the name of Mats Gustafsson who had just graduated with his PhD from John Clarke's group at UC Berkeley working on low-temperature AFM. Mats knew little about optics and was actually not keen on trying interferometry, but he agreed to attend an astronomy conference held on the big island in Hawaii. He returned enthused (Hawaii can do that), and Gustafsson dug into the idea of expanding interferometric microscopy into the far field.

Over the next several years, Gustafsson advanced the ideas into a succession of different types of interference microscopes. First there was I^2M (Image Interference Microscopy) that used two opposed microscope objectives to collect emitted light and superpose the two signals interferometrically onto a camera. This configuration was similar to a method developed by Stefan Hell and Ernst Stelzer in Heidelberg, Germany, called 4Pi microscopy[16], illustrated in Figure. 7.4. The I^2M mode is to 4Pi

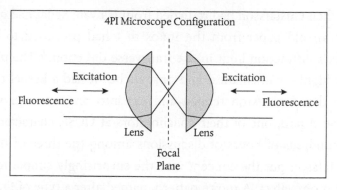

Figure 7.4 Configuration of the 4Pi confocal microscope. The excitation is coherently superposed at the sample on the focal plane, and the fluorescence from the sample is coherently superposed on a detector. The focal volume size is limited by the Abbe resolution for transverse positions but has enhanced axial resolution because of the standing wave interference.

microscopy as the Fizeau configuration of stellar interferometry is to the Michelson configuration. I²M was detected in the image domain while 4Pi microscopy was a point scanning technique. Gustafsson then moved on to I³M (Incoherent Illumination Interference Microscopy) that takes an incoherent light source and illuminates the sample through both objectives, selecting zero path difference to superpose the short-coherence light interferometrically. When I²M was combined with I³M, it gave I⁵M. Combining interferometric illumination with interferometric detection, the I⁵M mode provides the highest form of axial resolution. The team found that the axial resolution in this far-field technique approached 70 nm.[17] However, this impressive resolution length along the optic axis was considerably smaller than the transverse resolution, which was still stuck at the Abbe limit, but even that was about to fall.

When Gustafsson had returned from Hawaii, Sedat had given him an old paper from the 1960s that had proposed to beat Abbe's diffraction limit in the transverse direction.[18] The paper was highly mathematical and opaque, but it had a kernel of an idea that Gustafsson refined and put into practice after what David Agard, one of their collaborators at UCSF, characterized as hundreds of hours of discussions among the three of them. Gustafsson put the concept into the surprisingly simple terms of a moiré effect. A moiré pattern, named after a type of French fabric, is a common optical illusion seen when two patterns, having a similar periodicity, are superposed on each other. For instance, looking through two layers of screens on a screened-in porch shows this effect very easily. It is caused by spatial frequencies that periodically overlap each other to produce dark bands interspersed by transparent bands.

Gustafsson applied the moiré effect to the microscope and came up with something surprising. If one took two transparent screens or periodic patterns that could *not* be resolved by the system (their feature sizes are below the Abbe detection limit) and lay them on top of each, then the resulting low-frequency Moiré pattern *could* be resolved by the microscope! In interferometry, this is called a heterodyne technique: taking two high-frequency signals and causing them to interfere to produce a low-frequency difference signal that is easily detected. In the case of microscope imaging, the heterodyne is entirely spatial, mixing high spatial frequencies in a target with a high-frequency carrier wave and producing a low-spatial-frequency signal that can be imaged. Because the carrier wave is a periodic pattern of bright and dark fringes, Gustafsson called the new technique structured illumination microscopy (SIM),[19] beating the Abbe diffraction limit

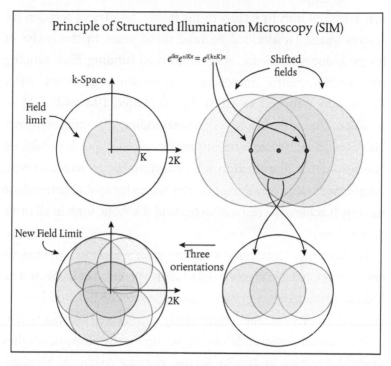

Principle of Structured Illumination Microscopy (SIM)

Figure 7.5 The principle of structured illumination microscopy (SIM) illustrated in Fourier frequency space. The conventional receptive field in microscopy is represented by a disc of radius K. By adding a spatial periodicity of spatial frequency K to the object, the disc is shifted laterally a distance K. After combining these shifts from a number of different orientations, a new receptive field is constructed (computationally) that has twice the spatial frequency for double the spatial resolution on the reconstructed image.

by a factor of two. The physical principle of SIM is illustrated in Figure. 7.5.

It is well known that post-doc positions in biology can go on for years, but Gustafsson's time as a post-doc was longer than most. As a student at Berkeley, he had taken a scholarship that required him to return to his home country of Sweden for several years. He decided to take these years in the midst of his post-doc with Sedat, who continued funding him, sending him interferometric microscopes to continue the work. After Gustafsson returned to UCSF he combined I^5M with SIM to produce the even more cryptic-sounding I^5S microscope.[20] This was the ultimate interferometric microscope! It combined interferometric illumination with interferometric detection with interferometric generation of carrier waves for spatial heterodyne mixing. It achieved a resolution beyond the Abbe limit in all three dimensions.

By this time, Gustafsson's talents were widely acknowledged and appreciated. He was offered a faculty position at UCSF, but he had no desire to write grants, which provide the life blood of any research faculty member. Fortunately, he received an offer to join a new top-tier research lab being set up by the Howard Hughes Medical Institute at Janelia Farms, outside Ashburn, Virginia. Today, Janelia is to biological sciences research what Bell Labs had been to physical sciences research before the mid-1980s—all the top researchers in the field, all in one location, just a short walk away down any hallway. The attraction to work at such a place was too good to refuse, and Gustafsson was wooed away from UCSF in 2008 to start up a super-resolution research group at Janelia. He began a collaboration with Eric Betzig who was pursuing an alternative approach to super-resolution microscopy based on single-molecule fluorescence. Tragically, Gustafsson was diagnosed with brain cancer to which he succumbed in 2011. Eric Betzig went on to share the 2014 Nobel Prize in Chemistry for super-resolution microscopy with Stefan Hell of Heidelberg University and W. E. Moerner of Stanford University, but Gustafsson was remembered and honored in the event at Stockholm. In his extensive Nobel autobiography, Stefan Hell says of Gustafsson and his contributions to the field of super-resolution microscopy, "He was of historic caliber."[21]

8

Holographic Dreams of Princess Leia

Crossing Beams

Help me, Obi-wan Kenobi. You're my only hope.

Carrie Fisher, *Star Wars* (1977)

In June 1977 a million moviegoers watched a universe come to life in a galaxy far, far away, where light and dark faced off in universal conflict, space cruisers jumped to light speed through dark space, and light-saber beams clashed in showers of sparks in George Lucas' epic film *Star Wars*. We watched Luke Skywalker clean the carbon-scored astro-droid unit R2D2 when he accidentally triggered a holographic projection—Princess Leia Organa calling for help from the Jedi knight Obi-wan Kenobi—a vision ephemeral and convincing. But it was all just movie magic, special effects added after the fact. At the time, popular holograms at novelty shops like Spencer's Gifts were impressive, but there was nothing that could project into three-dimensional space in real time. The hologram of Princess Leia planted the seed of 3D holographic videos into the minds of a generation.

Interference. David D. Nolte, Oxford University Press. © David D. Nolte (2023).
DOI: 10.1093/oso/9780192869760.003.0008

The Holographic White Elephant

The inventor of holography spent 10 years of his research life driving toward a goal he never reached. Grants were awarded, research was undertaken, intriguing half-results were obtained, a few brilliant scientists were seduced, and then the field died. But just as its last embers were extinguished, it resurrected, like the Phoenix, in the hands of an entirely new set of scientists studying entirely different things. Holography was reborn as a soaring success, catapulting its inventor, Denis Gabor, to the Nobel Prize in 1971 long after he had left the field behind.[1]

Denis Gabor (1900–1976) was born at the turn of the century just as physics began a revolutionary change. His small neighborhood of Budapest, Hungary, produced the likes of Theodore von Karman, John von Neumann, Leo Szilard, Edward Teller, and Eugene Wigner, all born within a few decades of each other and all within a few city blocks. Gabor was an astute student, and his degrees in electrical engineering landed him a job at the Siemens electronics company on the outskirts of Berlin. Always curious, he attended university lectures given by Einstein, Planck, Nernst, and von Laue. But Hitler's rise to power in 1933 sent him into exile to England where he found work at the research lab of an electronics company known as BTH. The war years separated him from his research colleagues after he was designated an Enemy Alien when he was isolated in a small shack outside the security perimeter. Ironically, this segregation also gave him complete freedom to work on long-term projects of his own interest away from the glare of supervisors grinding out work for the war effort. One of his many ideas was on super-resolution of the electron microscope.

Gabor was driven by a missed opportunity earlier in his life that still galled him. For his PhD thesis, he had developed a system with a high-voltage electron beam that he steered by a rudimentary magnetic lens he made. Shortly after completing this work and going to Siemens, Hans Busch, then a student at the University of Jena, perfected Gabor's magnetic lens, which was used a few years later by Ernst Ruska and Max Knoll in their invention of the electron microscope. Gabor, like most ambitious scientists, suffered mild delusions of grandeur, and when he thought about how close he had come to inventing the microscope himself, he knew he had missed his big chance to invent the electron microscope. Isolated in his hut at BTH, he began thinking about the resolution of current electron microscopes that could see features down to about 5 nanometers. But based on the energy of the electrons and their de Broglie wavelength, the actual resolution should have been as small as 10 picometers! The main limitation of the resolution of the electron microscope was the aberration in the magnetic lens.

In the summer of 1947, he was sitting on a bench at a tennis club waiting for his match to begin, watching the tennis ball sail over the net in three-dimensional trajectories that always found a tennis racket at each end, like photons traveling through a system of lenses always ending at an image. Since his student days in Budapest, Gabor had been intrigued by the fact that Huygens' principle applied not just to an image on a screen, but also to an entire volumetric wavefield that filled space. Even if an image of an object is formed at one specific plane in space, like the face of the tennis racket, all the information must be present at all other points in space, contained in the volumetric wavefield as the partial waves of the object interfered with each other in 3D. Electron waves in an electron microscope would certainly do the same thing. His epiphany was realizing that a photographic recording of the electron interference pattern in an electron microscope would capture the full three-dimensional information of the wavefield. This recording could then be read out using optical means, using high-quality aberration-free lenses to reconstruct the object with the full resolution allowed by Abbe's diffraction limit for light.

Convinced that he could improve the resolution of electron microscopy by using this two-stage approach, he lobbied the BTH management for time and money to pursue a proof-of-principle demonstration. His managers were not supportive at first, because BTH was an electronics company, and Gabor's project was purely optical, but he was eventually given several months to test his ideas. The first stage, recording electron interference patterns, was too ambitious, but the second stage, reconstructing images by diffracting light off of a photographic film, could be done in a reasonable time. In July, Gabor began working with his assistant Ivor Williams on optical wavefront reconstruction. The requirements for coherent light were severe, forcing them to use a high-pressure mercury arc lamp to get high intensity with sharp spectral lines for temporal coherence, passing the light through a pinhole only three microns in diameter to give it spatial coherence. The remaining light intensity was too low to expose any large item, so they chose small objects of only one millimeter diameter to produce interference patterns on a piece of ultra-sensitive photographic film only 1 cm on a side. After the film was developed and illuminated from the same pinhole, the reconstructed object was so small it had to be observed through a microscope.

Despite the small sizes involved, the photographic patterns on the film were striking in their complexity, as Gabor had foreseen, caused by the complicated partial wave interferences from the illuminated object. Gabor called it a "hologram," from the Greek roots for *whole* and *picture*, because every region on the hologram contained the whole information of the picture. The truly amazing thing was that the seemingly meaningless blotches of dark exposed film could be used to reconstruct recognizable letters from the object. It was as if the hologram were a cryptographic code that only light could decode back into a clear image. The image, viewed from the far side of the hologram looking back at the light source, was a virtual image (see Figure. 8.1), so it had to be imaged using lenses onto a photographic plate in order to record the reconstruction. The letters he chose for his 1-mm object, which was created as a microphotograph, composed three names: Huygens, Young, and Fresnel.

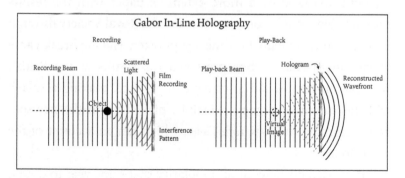

Figure 8.1 Gabor in-line holography has two stages. First, the light scattered from an object interferes with the incident beam, and the interference pattern is recorded on film. Second, the photograph (hologram) is illuminated by the incident beam, generating the reconstructed wavefront of a virtual image.

Once Gabor and Williams had successfully demonstrated optical wavefront reconstruction, Gabor began a publicity campaign, lobbying scientists wherever he could find them to see demonstrations of the effect in his laboratory at BTH. He approached Sir Lawrence Bragg, one of the most prominent figures in British science, to see if he would be willing to communicate his paper to the Royal Society. But Bragg was hesitant, not sure he understood the principles behind wavefront reconstruction. Despite Gabor's knack for promotion, he had buried the essence of wavefront reconstruction underneath complicated integrals that were difficult to follow. On the other hand, his laboratory demonstrations could not be resisted, and Bragg visited Gabor's laboratory to see the reconstruction with his own eyes. Seeing is believing, and shortly afterward Bragg wrote to Gabor to say, "I think I am beginning to understand the principle, though it is still rather a miracle to me that it should work."[2] By this time, Gabor had published a short paper in *Nature* and had filed for a patent, followed by a more extensive paper with the results shown in Figure. 8.2, communicated to the Royal Society through Bragg, who had finally become a supporter. The big break came in late 1948 when Gabor presented his results at the British Association for the Advancement of Science, making a splash with his new microscopy principle, picked up in the newspapers, including the *New York Times*, which ran an account of the invention.

Gabor used his newfound popularity that same year to obtain a position in electronics at Imperial College, where he continued his optical work on wavefront reconstruction, but he delegated the first-stage generation of the electron hologram to a sister

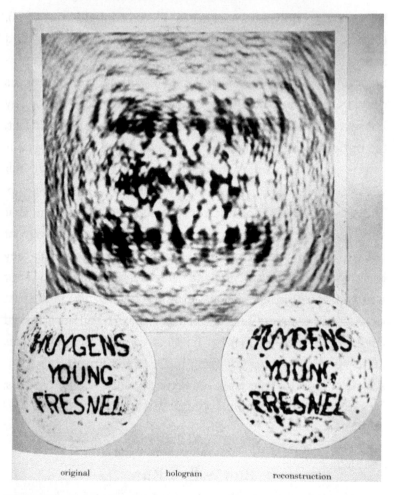

original   hologram   reconstruction

Figure 8.2 Images from Gabor's 1948 paper. The original (lower left) had a diameter of 1.5 mm. The hologram (center) was recorded 50 mm from the object. The reconstruction (lower right) from light scattered by the hologram was viewed (and re-photographed) through a microscope objective.

Credit: Royal Soc. of London.
D. Gabor, "Microscopy by Reconstructed Wave-fronts," *Proceedings of the Royal Society of London Series A—Mathematical and Physical Sciences*, vol. 197, no. 1051 (1949): p. 454.

lab of BTH known as Associated Electrical Industries (AEI). Unfortunately, the coherence of electron sources was not as good as hoped, and the AEI team had great difficulty creating electron holograms in the electron microscope. Although some fringes could be recorded, there were not nearly enough to provide any improvement to the original electron microscope images. In fact, the optical reconstructions from the paltry electron holograms had noticeably worse resolution than the original. This difficult work continued for 10 years until 1956, when AEI finally gave up. The two-stage diffraction microscope was a dead end. During that same time, several other groups had taken up optical wavefront reconstruction, but one-by-one these also had shut down as they began to view Gabor's subject as a white elephant.[3] By 1961, Gabor was wistfully writing of his wavefront reconstruction in the past tense as an interesting interlude of his career that had not achieved any of the hoped-for benefits for electron microscopy. He was done—but just at that moment, holographic wavefront recon-struction found a killer app, unrelated to electron microscopes, for classified Cold War military radar research.

Run Willow Run

On the outskirts of Detroit, Michigan, along the Huron River, a small nondescript regional airport sits at the edge of the old car-manufacturing town of Ypsilanti. The airport supports a number of regional airlines as well as the Yankee Air Museum that hosts air shows in the summer. It is also the site of the largest single-roof factory, now closed, that had built B24 *Liber-ator* bombers during WWII. On the east side of the runways, a number of buildings are still left standing from the Willow Run Laboratories that operated from 1951 to 1972 funded by Cold War military money and administrated by the University of Michigan in nearby Ann Arbor. These bare windowless buildings are where holography, unbeknown to Gabor at the time, gained its maturity in the hands of a talented young researcher just beginning his career.

Emmett Leith (1927–2005) was born in Detroit, Michigan, and never really left home. He received his bachelor's and master's degrees in physics from Wayne State University in downtown Detroit, then joined Willow Run Labs, 20 miles from home, starting at a yearly salary of $3,950 in 1952. When he joined the lab, radar technology was going through a major shift with the advent of side-looking radar which allowed an aircraft to generate a two-dimensional radar map of a large area to the side of the flight path. This was useful, for instance, when the aircraft could stay on one side of a country's border and make radar maps of the terrain on the other (enemy) side of the border, as if it were flying directly overhead. This type of radar mapping is known as synthetic aperture radar (SAR). In its earliest form, it relied on the Doppler effect on reflections from targets ahead of or behind the flight path. But the Doppler effect also made the radar data difficult to process at a time when computers were not widely available and were difficult to program, so the SAR researchers were looking for other, faster solutions.

One possibility for fast processing of SAR data was to take raw radar data, which were in a scrambled form of staggered echoes and Doppler frequency shifts, and record the data using optics onto photographic film. The film would then be illuminated with light and manipulated with lenses to form the direct image on the final recording plane. This two-step process—radar recorded onto film, then film optically imaged onto an image plane—was similar to the two-step process that Gabor had originally hoped to use with electron microscopes. Although the information from the electron microscopes had been too difficult to record, it was easy to record radar data onto film. Therefore, this highly specific application of radar gave birth to the first optical processor—an analog computer that used light instead of electrons to calculate.

Leith had taken several courses on physical optics for his physics degrees, and he was picked as one of the first team members on the optical processing project for SAR. The central operation needed for SAR image reconstruction was a Fourier transform, which is also the central function of a lens in an optical system, as illustrated in Figure. 8.3. For instance, when an object is placed at the front focal plane of a lens and illuminated by

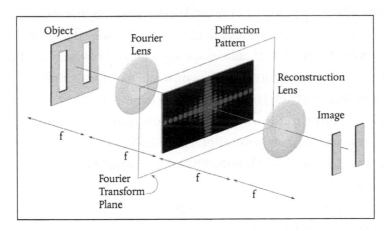

Figure 8.3 Optical Fourier Transform. When a lens is place one focal distance from the object, it physically generates a Fourier transform of the object on its back focal plane. This Fourier transform can be reconstructed back into the image domain by a second lens.

light, the optical field at the back plane of the lens is the Fourier transform of the object. Furthermore, it is easy to manipulate data in an optical system by simply putting masks in the optical train at select positions to act as spatial filters. In the 1950s, performing a filtered two-dimensional Fourier transform using an electronic computer would take hours or even days, but the lens system worked literally at the speed of light. The limiting delay in the whole process was the exposure time for the photographic film.

Leith began digging into the operation of optical processors and formulated a new way of thinking about information processing by optical systems. He viewed the system in terms of diffraction gratings, especially Fresnel zone plates, and how they affected coherent light. This way of thinking brought him to search the literature, and he found a paper, published in 1956 by a group at Stanford, on Gabor's wavefront reconstruction. From there Leith found the full literature published by Gabor as well as from the few labs that had pursued those ideas. This was exactly at the time when holography went into a hiatus after the failure of AEI to create workable electron microscope holograms. But in the secret world behind security clearance, this was the beginning of holography in a new venue. Leith's optical processor was using the radar data—translated to photographic emulsions—as holograms that were reconstructed using light to generate images of buildings and terrain.

By 1957 the SAR system was ready for field trials, but the first eight data-collection runs failed to produce holograms that could generate recognizable images when read out with coherent light.[4] The aircraft experienced too much buffeting by wind and turbulence which ruined the coherent relationships of one radar data set to the next. Only after they started flying on cloudless nights around 4 a.m. did they finally succeed in generating a clear field map from the SAR data. From that point onward, attention turned to stabilizing the flight path as they created ever-improving SAR images using the optical processor. By 1960, the optical processor was considered a great success, and the Army even risked posting a press release on the unprecedented capabilities of SAR, while keeping the optical aspect top-secret. No one outside knew, not even Gabor, that Willow Run was performing holography on a daily basis.

Ever since Leith had read the wavefront reconstruction paper in 1956, he had toyed with the idea of generating holograms from real-world objects, not just radar data. When a young new researcher joined his group in 1960, Leith had the time and manpower to pursue this idea seriously. Juris Upatnieks (1936–) was born in Latvia and emigrated with his family to the US in 1951, completing his bachelor's degree in electrical engineering from the University of Akron, Ohio, in 1960 and then joining Leith's optics group at Willow Run. By the end of the year, Leith and Upatnieks had reproduced Gabor's first hologram (the text "Huygens, Young, Fresnel") and were forging ahead with more ambitious plans. One of the most serious drawbacks of Gabor's holography was the presence of a "twin" out-of-focus image that overlapped and distorted the direct image in the reconstruction. This twin-image problem was a serious drawback of holography that no one had succeeded in solving—in the West. Behind the Iron Curtain, the Russian researcher Yuri Denisyuk (1927–2006) had made thick photographic emulsions (volume holograms) that had eliminated the twin, but his work was unknown at Willow Run.

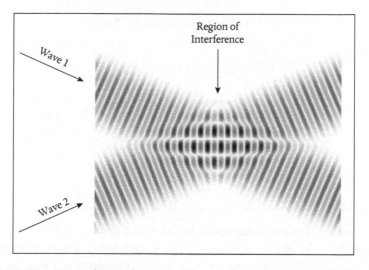

Figure 8.4 Two coherent beams crossing each other. The wavefronts propagate at the speed of light, but the horizontal interference fringes, where the waves overlap, are static.

To eliminate the twin while still using two-dimensional film, Leith and Upatnieks began experimenting with a diffraction grating to produce two separated beams that crossed at a small angle. This produced classic interference fringes, as shown in Figure. 8.4, that they recognized as a spatial carrier wave, which modulated the hologram like AM radio. When the recorded holographic grating was illuminated by coherent light, it generated an image in the first diffraction order that was free from the twin, as shown in Figure. 8.5, improving the quality of the reconstruction dramatically. Today, this technique is called "off-axis holography" which distinguishes it from the "in-line holography" used by Gabor. Leith and Upatnieks promptly wrote two papers, one on the carrier-wave off-axis solution to the twin-image problem,[5] and the other on wavefront reconstruction from the

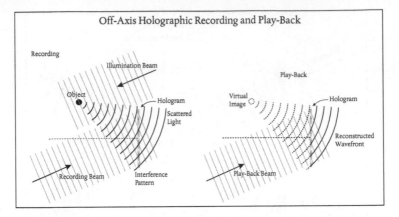

Figure 8.5 Off-axis holographic recording and play-back. The scattered light from the object interferes with the recording beam to expose the holographic film. After the film is developed, illuminating it with the play-back beam causes the beam to diffract in the form of the original wavefront.

physical-optics point of view of spatial carrier frequencies.[6] No sooner had Leith and Upatnieks obtained their best results in wavefront reconstruction, than Upatnieks left for military duty, and all progress on holography stopped—literally worldwide, because Gabor had left the field, as had the few researchers inspired by him, and even Denisyuk in Russia had turned to other interests. But in a research lab far out west, a lone researcher made a discovery, during the same month that Upatnieks had joined Willow Run, that would bring Leith and Upatnieks back to their holography lab with a powerful new tool.

The Light Fantastic

The invention of the laser is one of the great dramas of physics in the twentieth century[7] —which is no mean feat, because it was a century that also saw the development of relativity, quantum mechanics, the nuclear bomb, superconductivity, the transistor, and quarks. All of these discoveries and inventions have backstories replete with intrigue, superhuman efforts, missed opportunities, and serendipity. The stories share in common a small community with a heightened sense of urgency, anticipating that something giant was about to happen. These were times when individual researchers wanted not just to be part of the next big thing, but also to be "the one" to make it happen. Such times are rare in science. In my own scientific lifetime, I can think of one moment like this, when high-Tc superconductivity was discovered by Bednorz and Müller, and the "Woodstock of Physics" took place at the March Meeting of the American Physical Society in New York City in 1987. The hallways outside the presentation hall overflowed with a press of physicists who had been drawn to the event like the second coming.

The LASER (light amplification by stimulated emission of radiation) was the culmination of a 50-year journey that began when Einstein predicted the process of stimulated emission in 1916. In classic Einsteinian style, Einstein used simple ideas of transition rates and equilibrium (known as detailed balance) to show that an electron in an excited quantum state can be stimulated by a photon to emit an identical photon. At first, this idea seemed to have no applicable consequences, until Willis Lamb at Columbia University realized that nuclear magnetic resonance (NMR) could be used to manipulate nuclear spins that would affect atomic transitions in the radio-frequency range.[8] Using this idea, Edward Purcell and Ezra Pound, working at Harvard University in 1951, created a population inversion, where more atoms were in an excited state than the ground state, as shown

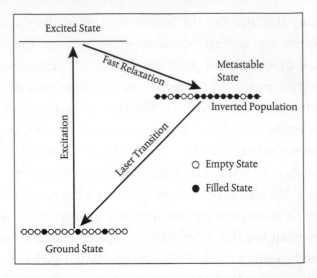

Figure 8.6 Population inversion in a three-level system. Atoms in the ground state are excited to the upper state and relax quickly to the long-lived metastable state where the population builds up, creating a population inversion. Stimulated emission brings the atoms back to their ground state.

in Figure. 8.6. They called it "negative temperature" because of the way that temperature is usually defined in terms of ratios of state occupations in equilibrium.[9] More significantly, in this condition they could demonstrate stimulated emission, where a single photon would trigger a chain reaction as one photon stimulated another, then the two photons stimulated two more, then four more, and so on. The effect decays rapidly, but Charles Townes (1915–2015), working at Columbia University, realized that stimulated emission could be maintained in a steady state if the population inversion were placed inside a resonant cavity. Furthermore, the radiation that leaked out of the resonator would be coherent, just like a radio antenna but at frequencies much higher than any radio could operate. Working with his student James Gordon (1928–2013), they demonstrated microwave amplification by stimulated emission, called the MASER, in 1954. Once the maser was established, it was a fairly obvious question to ask whether a system might be created that could operate at optical wavelengths.

The race for the laser began in 1956 when Townes sat down to talk with Gordon Gould (1920–2005), a 37-year-old graduate student at Columbia University. Each of them had an agenda. Townes was looking for ways to shift maser operation from microwaves to the optical range of frequencies, and he knew that Gould was studying optical pumping for his doctoral research, which Townes thought might be a route to population inversion. Gould, on his part, was thinking how to get light to amplify through stimulated emission, and Townes was the expert on stimulated emission. After two discussions, they each went their own way. Gould wrote down a number of prescient ideas in a notebook that he had witnessed and notarized, and which became the centerpiece for many decades of legal battles. He coined the name "laser" for his invention. Townes teamed up with his bother-in-law, Art Schawlow (1921–1999), an expert on spectroscopy, to work out the theory of what they called the optical maser, publishing their work in the December 1958 issue of the *Physical Review*.[10]

One of the problems that both Gould and Townes solved was how to map the operational parts of the maser onto the laser. The maser operated by putting molecules inside a radio-frequency cavity whose size was comparable to the radio-frequency wavelength and would resonate at some small multiple of its fundamental half-wavelength mode. The first maser based on ammonia molecules operated at a microwave wavelength of

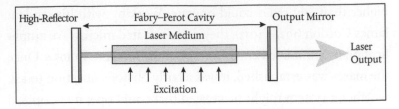

Figure 8.7 Fabry–Perot Laser Cavity. The beam bounces back and forth between the mirrors, amplified on each pass by the population inversion inside the laser medium. A small fraction leaks out the front mirror to form the external laser beam that is bright and collimated.

12.5 cm, which was a convenient size to construct a microwave resonator. However, light wavelengths are almost a million times smaller, around a micrometer in length, and there was no practical way (at that time before nanoengineering) to create such a small resonant cavity. The Fabry–Perot interferometer, on the other hand, with lengths of many centimeters, is an optical resonator that operates at extremely high multiples (in the millions) of its fundamental mode. Fabry–Perot etalons are constructed of two mirrors facing each other, shown in Figure. 8.7, held rigidly to within a tiny fraction of a wavelength, where there is plenty of space in between the mirrors to place the population-inverted medium.

Gould dropped out of graduate school to try to build the first laser. Having no funds of his own, he joined a company that agreed to find grant money to support the laser development. Unfortunately, the grant came from the military, and Gould had been a communist sympathizer (he had earlier been dismissed from the Manhattan Project for that reason), which precluded him from working directly on the project. Meanwhile, ideas began to fly around in the budding laser community. No one knew what kind of material could be induced to amplify stimulated emission of light. Some thought the excitable medium should be a gas, others a solid. Some thought that optical pumping could produce a population inversion, others thought electrical stimulation might work. In the midst of all the ideas and uncertainty, one of the many contenders racing to make the first laser who was working at a little-known corporate research lab in California decided to try pink ruby.

Hughes Research Lab (HRL) is perched high above the exclusive beaches of Malibu on a rare stretch of the California coast that faces due south. This may be part of the allure of Malibu beaches that catch more sun than any other beach in the state. Standing on the shore at the height of winter, you can watch the sun rise and set over water. In the spring of 1999, I visited HRL for the first time. As I waited in the lobby for my host, David Pepper, to escort me back to the labs, I noticed a large display case holding a gadget that looked like something out of a cheap Buck Rogers movie. When I looked closer, I realized with momentary awe that it was the first laser, followed by disappointment when I realized it was just a replica. The real one is stored away for safe keeping, although it was loaned out in 2010 for a public demonstration (and it still worked!) at Simon Frasier University in Vancouver, Canada, to honor the man who built it and made it work 50 years earlier—Theodore Maiman.

Theodore Maiman (1927–2007) returned from the last days of WWII as a radar expert who would rather be a physicist. He applied several times to the Stanford physics department but was turned down. On a whim, he hitchhiked from Denver to Stanford and talked his way into their master's program in electrical engineering. After obtaining his degree, the physics department finally admitted him into their PhD program, and he began working with

Willis Lamb, who had just transferred from Columbia and needed someone to set up his new lab. After Maiman filed his thesis in 1955, Lamb asked him to stay on in his group, but Maiman opted instead to spend his life's savings on a cruise around the world. This may not have been a great career choice, because Lamb was awarded the Nobel Prize in Physics later that year. Nonetheless, Maiman landed a plumb research position at HRL where he was put in charge of improving ruby masers for a military contract. The ruby rod he used for his maser was known as "pink ruby" for its light-pink color caused by chromium impurities in the crystalline lattice.

Maiman's success on the project brought him an invitation to talk at a conference on "Quantum Electronics—Resonance Phenomena" held at Shawanga Lodge in the Catskills in September 1959, organized by Townes. There were only 66 speakers, and virtually all the talks were on masers, but two key talks introduced the optical maser—one by Art Schawlow and the other by Ali Javan (1926–2016)—both from Bell Labs. Javan described his progress on a gas laser based on helium and neon atoms (HeNe) in a plasma. Schawlow gave a theoretical talk in which he analyzed the likelihood for population inversion and stimulated emission in various materials, including pink ruby, which he concluded would not work for an optical maser. Despite the conference emphasis on microwave masers, the talk in the hallways was on optical masers, and by the time the conference was over, the attendees had the strong impression that the first optical maser was just months away, most likely coming from Javan's efforts on the HeNe system at Bell Labs. Maiman felt differently.

As soon as he returned to HRL, Maiman asked his manager's for permission to try pink ruby as an optical maser. Because of his extensive experience with the pink ruby, Maiman believed that Schawlow's analysis had missed an essential point. Furthermore, Maiman was particularly adept at building novel equipment, so he felt he had a shot. The Hughes management reluctantly agreed, and Maiman began an intense sprint to try to beat Javan at Bell Labs. Over the next months, he and his technician Irnee D'Haenens explored the optical properties of the pink ruby and designed laser cavities and pump lamps. One of Maiman's key decisions was to use an intense optical excitation from what was essentially a very powerful flash bulb that was wrapped around the ruby rod. If the ruby amplified stimulated emission, it would be very brief, but he knew the spectral signatures to look for that would distinguish laser light from the ordinary background red fluorescence that ruby always displayed.

The critical moment came on May 16, 1960, not quite seven months after the Shawanga Lodge conference. Maiman and D'Haenens, working alone, slowly increased the voltage on the flash lamp, until suddenly the room was bathed in red light. D'Haenens was color blind for the color red, but the ruby red laser light was so bright that even the tiny fraction of red receptors in his eyes responded to the flashes, and D'Haenens saw the color red for the first time in his life—it was almost a religious experience for him.[11] Excited at their success, they called in researchers from neighboring labs to experience the first light of the laser. Later that night, word spread through the labs, and Maiman was met the next morning by his manager who wanted to issue an immediate press release. Maiman wouldn't have it. He insisted on testing the system carefully and writing a thorough paper for publication. Unfortunately, he chose the journal *Physical Review Letters*, which had an authoritarian attitude about acceptance criteria, and the paper was rejected. Not wishing to lose the race to Bell Labs, Hughes had Maiman rework the paper into a short note to *Nature*, which immediately accepted it for publication. The press conference was held on July 7 in New York City.

Tiny Hughes Research Lab, with far less time and money invested, had beaten vaunted Bell Labs to the goal. David versus Goliath. But gloating was not a good idea, because Bell Labs had its finger on the pulse of scientific power. Within five months, Ali Javan had his HeHe laser working in continuous mode in the infrared, rather than flashed, and his paper was accepted immediately for publication in *Physical Review Letters*. But neither he nor Maiman, nor several others in those first months of laser firsts, ever received the Nobel Prize. That went to Townes in 1964 and to Schawlow in 1981 for the conceptual basis of the laser—not the invention itself.

Laser development progressed at nearly warp speed through the early 1960s, expanding the types of materials that could be used and the ways they could be pumped to create population inversions. Dozens of different types of lasers had been invented just by 1965. Most of them were hard to operate, or were too inefficient to be useful, or worked at invisible wavelengths that made them difficult and dangerous. But some showed promise for applications. One of them was an extension of Javan's HeNe laser that had originally operated in the infrared at a wavelength of 1.15 microns. In 1962, Alan White and Dane Ridgen at Bell Labs operated a continuous-wave HeNe laser at a wavelength of 632.8

nanometers in the red as the first continuous visible laser. The red HeNe laser was easy to operate and could be made cheaply for general use in laboratories, and it was one of the first to be a commercial success. At the end of 1962, Upatnieks had returned from military service, and he and Leith bought a red HeNe laser for their holography lab at Willow Run.

Into the Third Dimension

In the background of all the early work on wavefront reconstruction was an implicit property that everyone sensed but could not quite put into focus, like an object in your peripheral vision on the edge of perception. That implicit property was the third dimension. Gabor was aware of it, and Denisyuk in the Soviet Union had exposed three-dimensional emulsions, while Leith and Upatnieks had perfected the reconstruction of three-dimensional wavefronts. All of the early work took advantage of the third dimension, but none of the practitioners realized its potential— until Leith and Upatnieks placed their first HeNe laser in the lab. With the high brightness and the long coherence of the laser, everything about holography changed. They could spread out the laser illumination into broadly diverging cones of light that could envelope both the object and large-format photographic plates. Before the laser, they had to use lenses to corral and collect the paltry weak light of their mercury arc lamp to concentrate their precious photons. Now, they could let the light run free over their lens-less optical systems as they made lens-less photographs of increasingly complex objects. Finally, those objects became three-dimensional.

The conversion from two-dimensional objects to three-dimensional was a slow and painstaking process. Despite the obvious advantages of laser light, it also came with strong negatives. The long coherence—the first laser they used had a coherence length of 25 cm—made every stray reflection in the optical system a source of interference, producing ugly blotches and bull's-eyes on the photographic film. Furthermore, they had to contend with laser speckle that buried images underneath tiny bright and dark patches of random constructive and destructive interference. Yet, as they confronted the challenges of coherent imaging for the first time, they also were finding solutions and work-arounds that have become standard procedures in coherent imaging labs around the world.

From December 1962 to December 1963, Leith and Upatnieks fought these battles in their spare time as they fulfilled their responsibilities to the many funded projects at Willow Run, most of them of military character, including continued work on optical processing for SAR. Finally, three days before Christmas in 1963, they made a high-resolution three-dimensional hologram of a toy train. The exposure took 10 minutes because of the diffusely spread laser light, and the train had to be weighed down with epoxy, inside and out, so that it would not move during the long exposure.[12] The photographic plate was large enough that one could look through it, using both eyes, at the diffused laser light illuminating it from behind. Magically, the toy train could be seen in full detail.[13] Most dramatically, it had the properties of parallax when one moved one's head, with the need to refocus the eyes when looking at parts at different depths, and it triggered binocular depth perception. Upatnieks later recalled that "... we were fascinated by the reality of the image and spent hours looking at it, and showed it to our colleagues."[14] Leith remembered that, "… people in the laboratory looked at it in astonishment, the management came in and looked at it, and the Director came in …."[15]

The final session at the spring meeting of the Optical Society of America, on Friday April 3, 1964, held at the Sheraton Hotel in Washington, DC, was on "Information Handling by Optics." Upatnieks gave the talk for the Willow Run group, and he invited the audience to visit the hospitality suite hosted by the laser manufacturer Spectra-Physics to see the latest best holographic reconstruction that they had made of the toy train. Later that evening, when Leith and Upatnieks went to visit the suite to see how it was going, they were surprised to find a long line of scientists trailing down the hallway and around the corner. Although it was a conference on optics attended by optical experts, there was a constant stream of disbelief as one scientist or engineer after another stepped up to view the holographic reconstruction. Almost no one, not even the experts, had ever conceived of such faithful capture of the full three-dimensional wavefield generated by light scattered from objects. It was especially difficult to understand how a two-dimensional recording on the photographic film could carry all of the three-dimensional information of the object—to many, it seemed like it was violating principles of information theory and dimensional invariance, but of course it was not.

The popular press was also captivated by the magical reality of the train that was not there. In the weeks following the conference, Leith and Upatnieks were inundated by phone calls and requests for interviews. Presentations at other conferences followed, such as a famous presentation in Boston later that year that was reviewed by a Boston reporter who encapsulated many of the unintuitive properties of holograms: "Because there are no lenses, each point on the object is recorded all over the photographic plate. So you can take a hologram and cut it in half—or in a dozen pieces—and each piece will still show the entire object, from a slightly different point of view, with only a little loss in sharpness."[16] Here was one of the first clear descriptions of the holistic properties of holograms. General distribution to the wider public came in 1965 as an article by Leith and Upatnieks in *Scientific American* introduced holography to a broad readership.

Within the technical optics crowd, one new feature of the hologram that attracted attention was the idea of the hologram as a window with a memory. It literally was storing three-dimensional optical information of an object, frozen in time, in a simple two-dimensional format. As a memory device, it represented considerable data compression combined with high storage density. Furthermore, if a two-dimensional memory could contain that much information, then what would be possible with three-dimensional optical memory? This simple question captured the imagination of a researcher at General Electric's research laboratory in Schenectady, New York. Unfortunately, it also got him fired.

Pieter J. van Heerden (1915–1999) was born and educated in Utrecht, The Netherlands, receiving his PhD in physics from Utrecht University in 1945. Shortly after graduating, he accepted an offer to be an assistant professor at Harvard, emigrating to the United States in 1948. That fall, as he was settling in to his new surroundings, he read Gabor's article in *Nature* introducing the concept of holography. He rushed into the office of Julian Schwinger (who would receive the Nobel Prize in Physics in 1965) proclaiming that Gabor had solved the phase reconstruction problem for light. Schwinger chided his overly excited junior colleague, saying that such a thing was known to be impossible. Schwinger suggested that van Heerden read the article again to find his mistake. When van Heerden persisted, telling Schwinger that Gabor "shows pictures," Schwinger read the article himself and was won over.[17]

An assistant professorship at Harvard is often a terminal position, though it is a nice springboard to a fruitful career elsewhere, and van Heerden switched to a research position at General Electric (GE) in 1953. Although his research was only peripherally involved with coherent optics—he was focused mainly on aspects of information theory—Gabor's paper made a lasting impression on him, simmering in the background of his thoughts, until the breakthrough of the laser brought it back in full force. Unaware of the work by Denisyuk and of Leith, van Heerden independently recognized the possibility of using laser light to store information as interference patterns in three-dimensional crystals. He wrote up a detailed proposal on a broad range of applications that included the storage of many overlapping images, providing the first analytic analysis of three-dimensional storage capacity using light. His first estimate on the storage capacity of a cubic centimeter crystal was an astounding 10 terabits.

Today, we have become inured to 4 and 8 terabyte drives, but in the early 1960s such a storage capacity was entirely impossible to believe. Van Heerden's manager was so skeptical that he quietly asked one of the other scientists in the department to review van Heerden's proposal. The report came back negative. This "failure" combined with van Heerden's penchant for promoting wild theories, such as "holographic" models of brain function, put van Heerden's future at GE in jeopardy. Fortunately, Edwin Land, the founder of Polaroid, had a more open mind and spirited van Heerden away from Schenectady to the Polaroid photographic film research lab in Boston. Photographic film is optical memory, and van Heerden's interests in three-dimensional optical storage were right at home. At Polaroid, van Heerden had the time and freedom to finish his analysis of three-dimensional storage, submitting a landmark paper[18] in the summer of 1962 before Leith and Upatnieks had purchased their first laser.

Van Heerden's paper was prescient for several reasons. Not only did it outline the first estimates on three-dimensional storage density, but it also recognized the holistic abilities of holograms that even today make them seem semi-magical. In particular, van Heerden explained how holographic memory could be associative. Associative memory is what happens when we are shown pairs of pictures that we commit to memory. Later, when we are shown one of the pictures, we remember its pair. Van Heerden realized that three-dimensional holographic memory would behave the same way. If the two beams writing a hologram carry two images, then when the hologram is illuminated later by one of the images, the other image is reconstructed, as in Figure. 8.8. This associative holographic memory seems almost mind-like in how it behaves, touching on the fields of artificial intelligence and neural networks. In van Heerden's paper there is even a drawing of a rudimentary neural network, obeying Huygens' principle, as he made an explicit analogy between holography and the brain. Reading the paper today, one gets the impression that van Heerden was 10 years or so ahead of his time. But that is where the idea stayed. As van Heerden shifted his attention more fully to aspects of artificial intelligence, the vanguard in three-dimensional

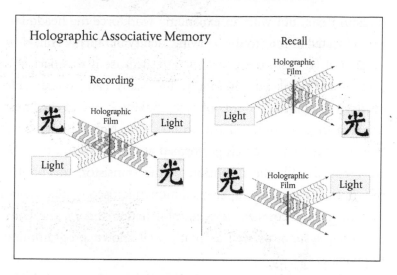

Figure 8.8 Associative memory as an optical "translator." A hologram is written using "light" and "hikari" (the symbol for light). When only one of these is later used to illuminate the hologram, the light diffraction reconstructs the other.

storage research shifted to the premier industrial research lab in the country, located in a quiet New Jersey suburb of New York.

Holograms on the Move

Bell Telephone Laboratories was founded in 1925 as a joint venture between Western Electric and American Telegraph and Telephone Company to pursue fundamental research in communications and allied sciences that could lead to discoveries and inventions serving the communications conglomerate. It was originally headquartered downtown in New York City in an office building between the West Village and the

Hudson River, but with an expanding workforce the headquarters relocated in 1967 to the growing facility in Murray Hill, New Jersey, not far from the Great Swamp. Because it was part of a government-regulated monopoly, Bell Labs had considerable freedom to operate and lots of cash, making it one of the finest research organizations in the world. Nine Nobel Prizes have been awarded for research performed at Bell Labs, and the labs are credited with the invention of the transistor, the photocell, optical tweezers, super-resolution microscopy, the digital camera (charge-coupled device), information theory, the Unix operating system, as well as many well-known programming languages.

The first laser had very nearly been invented at Bell Labs by Ali Javan, before being scooped by Maiman at Hughes, but many of the succeeding laser systems were invented there. The laser promised to be an entirely new communication technology based on light rather than on electric currents in telephone wires. As soon as researchers at Bell Labs learned of Maiman's ruby laser, they made one themselves and hauled it to the top of an old radar tower in Murray Hill and aimed it at another tower 25 miles away at Crawford Hill where observers could see the dim red flashes.[19] This dramatic demonstration of free-space optical communication made a nice press release for Bell and helped steal some of the thunder from Hughes.

Part of communication is control of information, and the growth of optical communication drove the search for the control of light by light. This is not simple, because light beams pass through each other without interacting, in contrast to electrons that draw on their electric charge and Coulomb effects to allow currents to control currents. However, it is possible to use high light intensities to modify the optical properties of crystals,

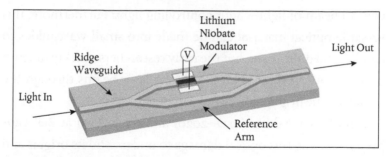

Figure 8.9 Lithium niobite Mach–Zehnder modulator in a ridge waveguide. Light entering the photonic integrated chip is split by the "Y" junction into a reference and signal arm. A voltage applied to the lithium niobite in the signal arm changes the refractive index and hence the optical phase. When the light is recombined in the second Y junction, destructive or constructive interference creates low or high intensities.

affecting how another light beam propagates through the crystal. This is called nonlinear optics, and it is one path for optical control. The development of the laser made high-intensity light beams a common resource, and the laser labs at Bell Labs began studying nonlinear optical effects in a broad array of new optical materials.

One of those materials was lithium niobate ($LiNbO_3$), a transparent colorless crystal that has optical birefringence and a large electro-optic effect in which an applied voltage changes the refractive index of the crystal. Such an electro-optic crystal can be used to make an electronically addressed optical switch by placing it in one arm of a Mach–Zehnder interferometer and applying a voltage, switching the light emission from one output port to another by changing the refractive index, shown in Figure. 8.9. Lithium niobate is also a nonlinear optical crystal that can be used in the same Mach–Zehnder to make an optically addressed switch by changing the refractive index with an intense control beam of light—light controlling light! Furthermore, this versatile optical material can be made into small waveguides to confine and direct light, but the early researchers quickly ran into problems when shining high-intensity laser beams through the crystals—the bright light damaged them.

In 1966, Art Ashkin (1922–2020) and Gary Boyd at Bell Labs began studying this optical damage to understand its origins and to possibly find ways to protect the crystals. They shone beams of visible and infrared light through small crystals and found they could image the track of the beams after the beams were turned off by viewing the crystals in polarized light. These results, plus observations of the self-deflection of the laser beam over time as it exposed the crystal, immediately suggested that the laser beam was modifying the refractive index of the material. Lithium niobate was known to be electro-optic, so they thought that the laser beam might be creating semi-permanent internal electric fields that remained after the laser was turned off. The crucial point of their work was that the photo-induced refractive index change, subsequently named the photorefractive effect, could remain for days or months, although shining other wavelengths of light on the crystal or heating it could cause the optical damage to relax. All in all, they viewed the effect as "highly detrimental" for the uses of the crystals in nonlinear devices.[20]

At that time, in the heyday of Bell Labs, internal research had an unusual structure where virtually anyone could work on almost anything that interested them. You were free to walk into anyone's office anywhere in the facility to strike up a collaboration, even if the collaborators were from different business units. One of the central divisions inside Bell Labs was Area 10, doing fundamental research, versus Area 20, doing applications and product development. Yet there was an easy flow of interests and ideas between them. As the optical damage work by Ashkin became known more broadly within the Labs, two researchers from Area 20 connected the dots between the semi-permanent optical damage in lithium niobate and the advantages of three-dimensional holographic storage.

Fong Chen and John Lamacchia, with the full resources of Bell Labs at their disposal, built an ingenious optical mixing experiment using two of the latest gas lasers—a HeHe laser plus an argon ion laser—both targeted at a lithium niobate crystal at the center of the setup. They used the argon laser to write holographic gratings inside the centimeter-thick crystal, relying on the optical damage effect of Ashkin and Boyd. The grating was read out by aiming the HeNe laser at the Bragg angle, shown in Figure. 8.10, and the diffraction efficiency was measured to be as high as 40 percent. Most impressively, in a later experiment, they directed the two argon laser write beams into opposite sides of the crystal to write gratings with extremely small fringe spacings of only 100 nanometers, and a diffracted signal was still measurable.[21]

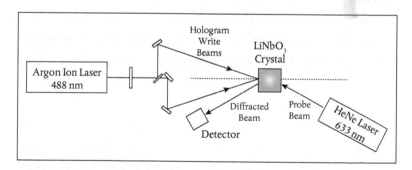

Figure 8.10 Chen and Lamacchia experiment writing gratings in a LiNbO$_3$ crystal with the argon laser and reading them out with the HeNe laser.

The experiment by Chen and Lamacchia was not the first time that three-dimensional holographic recordings had been made. Denisyuk in the Soviet Union and Emmet Leith at Michigan had been working with three-dimensional holograms in photographic emulsions. But the fringe spacings in the lithium niobate were far smaller than the smallest fringe spacings that were possible using emulsions and could support astronomically high data densities for three-dimensional data storage. Furthermore, the exposure times could be shorter than for photographic film, reducing the need for complicated vibration isolation. But many questions remained. In particular, the cause of the optically induced refractive index change remained elusive, as some crystals showed very large effects, while others showed no effect at all.

Around that time, a freshly minted PhD from the University of British Columbia joined Bell Labs in the Materials Science group. Alastair Glass (1940–) was given an empty lab, the resources to fill it, and a mandate to do something interesting—perhaps only marginally constrained by the mission of the group. Glass became aware of the work by Ashkin, Boyd, Chen, and Lamacchia. As was the norm for the Labs culture, where anyone could talk to anybody about anything, Alastair approached Ashkin to ask if he would like to collaborate to discover the origin of the optical damage. As was also normal for the Labs culture, Ashkin told him to get lost—they would do it themselves.[22] Egos could be large at Bell Labs, and it was not unusual for people to get a little short, especially if someone was homing in on their perceived turf. Ashkin never got back to it, because there was always something else coming down the road. In his case, it was optical levitation, work for which he received the Nobel Prize in 2018.

But just as anyone could tell another to get lost, anyone could also do what they wanted at Bell Labs, so Glass teamed up with a colleague, George Peterson, an expert in nuclear magnetic resonance, and they discovered that the damage was correlated with iron impurities in the crystals. The effect could be nearly eliminated, or else greatly increased, by appropriate control of the impurity content. Working with an out-of-date ruby laser that would only fire about one flash per minute, they illuminated single-crystal lithium niobate with the red ruby light and discovered that the flash produced a high-voltage spike across the crystal. In further experiments, after they got their hands on an argon ion laser with a continuous output and aimed it at the crystal, the voltage became so large that the crystal spit out electrical arcs—like a miniature version of something out of Nikolai Tesla's old lab. Such an effect could not be explained by any of the then-known photovoltaic effects[23] that always required junctions, interfaces, or chemical gradients. Glass and a post-doc from Germany, Dietrich von der Linde, proposed that the effect was due to the asymmetry of the $LiNbO_3$ crystal structure as light was absorbed by the iron impurities, giving an impulse to the photoexcited electrons along the polar axis. Evan Kane, a famous band structure expert down the hall, told Glass this was impossible because of transport symmetries, but Glass subsequently has speculated that this was the first demonstration of the transport of localized carriers, known as polarons, where the band structure does not participate. Regardless of whether the effect should exist or not, it made optically induced gratings stronger and faster and potentially more useful for light to control light.

As the new field of photorefractive effects grew, attracting interest from labs around the world, new types of materials were found that also had optically induced gratings, some of which were semiconductors with very fast write and erase times that could be used as optically addressed switches. The holograms in these materials were dynamic and could move and adjust to changes in the optical phases in the laser beams, opening up the field of dynamic holography and moving gratings. Adaptive interferometers became possible that could compensate for the types of mechanical motions that had plagued all interferometers since the days of Michelson, while performing exceedingly high-sensitivity interferometric detection of high-frequency surface displacements for ultrasound detection,[24] among other applications.

An explosion of volumetric holography techniques in the 1970s through the 1980s was driven by textbook examples of optical information processing in which photons in photonic devices behaved as electrons do in electronic devices. New and clever photorefractive applications seemed limitless, with entire conferences devoted to them. For example, optical neural network architectures were developed—van Heerden's pet idea—based on photorefractive effects, which had almost human-like tendencies to dream and to remember and to recognize.[25] In addition, data storage on magnetic discs was becoming a serious bottleneck for information systems, and the potential data densities of holographic memory seemed like a good solution, helping to launch holographic storage companies, like In-Phase Inc., spun out of Bell Labs in the late 1990s.

However, like many technological fields that wax and wane and wax again in cycles, holographic information processing was not the Holy Grail that many had hoped. When fighting against entrenched technology, like silicon chips or solid-state drives, it is hard to make inroads when the development time is too long and expensive, because the entrenched technology keeps making progress itself. Likewise, when fighting against digital storage, it is hard to beat the decreasing costs and increasing densities. But just as the 30-year Moore's law for electronics finally saturated, magnetic memory is also running out of space to store the exponentially increasing data on the World Wide Web. This may provide an opening for the next cycle of holographic memory and optical data applications. But in the meantime, various types of holograms have become routine elements of daily life as they contribute to our work and play.

Holograms at Work and Play

In April 2012, a quarter of a million concertgoers had their minds blown at the Coachella Music Festival at the Empire Polo Club in southern California, when the dead rapper Tupac Shakur, who had been shot and killed 16 years before in gang-related violence, appeared on stage, fully alive, singing rap medleys with the living rappers Snoop Dogg and Dr. Dre. He even talked to the crowd using the obligatory foul language of rapper shows. His image was life-sized and dynamic, roaming the full width of the stage. But it was just an illusion known as Pepper's Ghost, using ingenious glass reflections developed by the showman John Henry Pepper in 1862 for a live reading of Dickens' *The Haunted Man*. The effect has often been mistakenly referred to as Tupac's "hologram," but there was nothing holographic, that is to say interferometric, in the resurrection of Tupac. The words "holography" and "hologram" have become such cultural memes that almost any optical technology that can form three-dimensional images has appropriated those names.

There are many other examples of non-holographic 3D displays. These techniques attempt to make light emerge from point-like sources in a point cloud distributed in three-dimensional space. For instance, water mist can be used as a semitransparent volumetric medium to scatter light from a scanning laser beam. Alternatively, pulsed lasers can ionize the air itself within a volume, creating three-dimensional plasma light emission to act as the point cloud of light sources. Surprisingly, one of the oldest forms of three-dimensional display is also one of the best, proposed in 1908 by a French optical physicist the same year he won the Nobel Prize for a form of color photography based on interferometry.

Gabriel Lippmann (1845–1921) was one of those figures whose ideas seemed to be half a century ahead of their time—his novel uses of light look modern even by today's standards. He was the intellectual heir to Fizeau, with his deep comprehension of the coherence properties of light, and he was a master at geometric optics, his ideas spanning from the wave-like to the ray-like nature of light. His beginnings were modest, born in Luxembourg to a middle-class family that moved to Paris amid the turmoil of the revolutions of 1848. He failed his admission exams to become a teacher, so he settled on a lesser career in physics, eventually becoming a professor at the Sorbonne. His career overlapped with Fizeau, Mascart, Jamin, and Cornu, all of whom were in Paris around that time, representing a golden age of French optical physics at the end of the nineteenth century.

Lippmann produced the first color photographs using a principle that is similar to the volume holography that Denisyuk developed 50 years later. By placing a fine-grained photographic emulsion in contact with a liquid-mercury mirror, and illuminating it with light, standing waves were formed through interference inside the emulsion. The standing waves exposed the silver halide crystals at the antinodes but not the nodes, generating a Bragg diffraction grating for that specific color after the film was developed. To make a color photograph, the film was exposed by the light from a colored scene, and the different colors produced standing waves of different periodicities in the photographic emulsion, creating superposed Bragg reflectors for the different colors incident on the film at different locations. When the developed film was later illuminated by white light, the Bragg stacks reflected only the colors that exposed the various locations on the film, creating a color rendition of the original scene. The process of this interferometric form of color photography is illustrated in Figure. 8.11. The ingenious use of the interference properties of light is all the more impressive given the short coherence lengths of natural light, which is why the photographic emulsion needed to be in direct contact with the mirror. Denisyuk's later use of this principle was made significantly easier by the long coherence of the laser light sources he used.

The same year that Lippmann was awarded the Nobel Prize for his color plates, he proposed a process he called *integral photography* for three-dimensional display. Integral photography is the precursor to what is today called light-field imaging and display. Light-field cameras are already commercialized—these are the digital cameras that can refocus on different parts of a scene *after the photograph has already been taken*. It works by using an array of

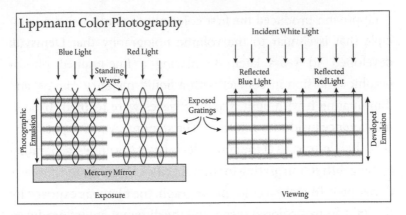

Figure 8.11 Principle of Lippmann color photography. Exposing a photographic emulsion on a mercury mirror creates standing waves that expose periodic layers depending on the color. When later developed and illuminated with white light for viewing, only the colors that had been present while exposing the gratings are reflected from those locations.

small lenses instead of a single objective lens. Each small lens sees a slightly different angle of the scene, and these small differences can be exploited using digital image algorithms to reconstruct the scene from different angles as well as from different focal depths. Lippmann proposed a similar procedure that can be used in reverse to generate 3D displays in which an image can appear to be floating in space before the viewer's eyes. In a two-step process, a scene is imaged by a lenslet array to expose photographic film, and then the developed film is illuminated with white light and viewed through the same set of lenslets. The original object is observed as three-dimensional, with properties of parallax and binocular vision.

With the advent of modern spatial light modulators (SLMs) and high-density light-emitting-diode (LED) arrays, combined with fast digital processing, light-field display is becoming the most advanced form of non-holographic 3D projection. The sizes of such displays are steadily increasing as are the angles of view. In the near future, these displays will become the primary way that consumers interact with three-dimensional objects—primarily as a novel means of advertising. Similarly, light-field headsets will gain traction through virtual reality (VR) and augmented reality (AR), mostly in the gaming marketplace. Although they are non-holographic, light-field displays are more advanced than current 3D VR goggles that simply present two slightly angled views of a 2D scene to the two eyes to trick the brain into thinking it is seeing a three-dimensional scene. But tricking the brain also brings with it serious eye strain and sometimes dizziness and nausea. Light-field displays do not trick the eyes because they generate light fields that closely mimic the actual way that light emerges from three-dimensional objects, removing most of the problems with visual discomfort. However, mimicking light fields is not as good as actually reproducing light fields, which still gives holographic display an advantage for some critical applications.

In simple terms, the difference between light-field displays and holographic displays is the difference between ray optics and wave optics. The lenslets of light-field displays direct rays of light to the eyes while a holographic display sculpts an actual wavefront like the real thing. Both types of display can use spatial light modulators, although light-field displays can also use dense arrays of LEDs, like those used in super-high-resolution LED monitors. But the primary difference is that the hologram shapes the actual wavefront, using coherent phase relationships, while the light-field directs dense sheafs of rays in different directions, as shown in Figure. 8.12. The profiles of the actual fields incident on

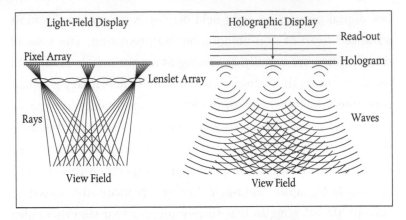

Figure 8.12 Light-field display compared to holographic display. Light-field technology directs rays of light through an array of small lenslets to mimic multiple point sources in space. Holograms reconstruct wavefronts that diverge from multiple point sources.

the eyes can be very different between the two display methods, which places more restrictions on the use of light-field systems than on holographic systems, but with relative trade-offs, because holographic systems are more difficult to manufacture and to operate, although they have higher fidelity.

Holographic displays are ideal for close-up interactions of an observer with a three-dimensional representation of an object because they control the visual wavefronts. The image can be made to hover in free space at arm's length in front of the observer so they can "touch" it and rotate it to look at it from any direction by using sensors that track the observer's hands and adjust the digital hologram in real time in response to the hand's motions. For instance, surgeons can use these systems to look at a 3D rendering of a patient's physiology to best plan the surgery. A brain surgeon can visually hold the brain of the patient in their hands, rotating it around this way and that to find the best incision angles. These systems come close to the famous scene in the 2008 Marvel *Iron Man* movie where Tony Stark is working with his digital assistant Jarvis to design his next-generation iron-man suit. However, in the fantasy world of the movie, Stark is using digital glasses, while in current real-world holographic displays, the user is constrained to stand or sit within a display unit that still has limited angle and range of view.

The differences between light-field and holographic display are currently disappearing as each borrows aspects from the other. Light-field displays are incorporating more aspects of phase control, getting better at shaping the actual wavefront, while holographic displays are borrowing the ease of ray manipulation using lenslets. At the same time, they both benefit from the shrinking pixel size in spatial displays, which further moves the technologies closer. The ability to manipulate more rays makes light-field displays more able to continuously shape the light fields, while the ability to manipulate finer spatial details make holographic displays less pixelated. The two technologies may soon merge into a composite form that uses the best advantages of both, or that continuously trades off one versus the other depending on the application. Large displays at a distance may use more aspects of light field, while high-resolution close-in displays may use more aspects of holography. But playing with holograms in entertainment and marketing is only one current trend in 3D display—holograms also do meaningful work.

When holography merged with digital cameras in the late 1990s, it created the new field of *digital holography* in which the interference fringes of the hologram are captured by the digital pixels, and image reconstruction is performed electronically by digital algorithms. In a broad sense, digital holography is synonymous with spatially resolved interferometry, performing interferometry simultaneously at many points in space with each pixel of the camera acting as the interferometric detector. As with physical holograms, the pixel-plane of the camera does not have to be on an image plane or even on a conjugate plane (a Fourier plane) but can be in a Fresnel configuration which is neither near-field nor far-field, creating a versatility for the optical design of holographic systems.

Many of the current applications for digital holography are for the measurement of small displacements, just as the Michelson interferometer and its variants are commonly used, but holography takes advantage of pixel arrays to measure displacements simultaneously across an entire spatial field. These interferometers are used for surface metrology; for instance, measuring surface displacements of mechanical parts that are subjected to stresses and strains or that are vibrating. When the target is composed of moving objects, small Doppler frequency shifts occur in the scattered light that cause the scattered wavefields to interfere and beat temporally against the stable reference field. By measuring the time dependence of the beat frequencies across spatial fields, it is possible to produce maps of motions, making movies of moving parts.

In the field of holographic medicine, coherent light scattered from biological samples is highly irregular, producing the spatial intensity noise known as *speckle* that was such a central player in the development of stellar interferometry (see Chapter 5). Holographic fringes modulating a speckle pattern are shown in Figure. 8.13. In a biological context, the bright and dark patches of speckle

Speckle Hologram

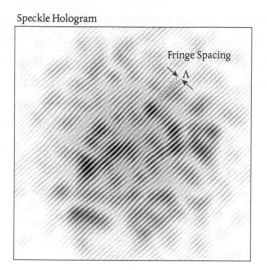

Figure 8.13 Speckle hologram showing interference fringes modulated by a complex speckle pattern such as from light scattering from a biological tissue.

caused by light scattered from living tissue behave as a large array of individual interferometers that sense the displacements of all the complex constituents of cellular matter. In living cells and tissues, these constituents are in ceaseless motion as the machinery of the cell performs its various functions. By measuring the shifts in intensity across the speckle pattern, and analyzing their time dependences, it is possible to extract internal speed distributions of the intracellular components,[26] and when therapeutics are applied to living tissues *in vitro*, changes in the intracellular Doppler spectra can be associated with the sensitivity or resistance of the tissue to chemotherapy.[27] By combining the Doppler spectroscopy with laser-ranging capabilities of holography, it is possible to create three-dimensional maps of tissue metabolism and drug responses.[28]

In many other medical applications, digital holography merges with the interferometric microscopes that were described in Chapter 7, such as phase-contrast imaging that uses spatial interference fringes to enhance sensitivity to phase shifts in single cells as well as performing quantitative phase imaging.[29] In the future, these applications will likely expand, using interference to provide unparalleled sensitivity to different biological functions and forms, enhancing our understanding of life processes while improving our ability to protect and preserve them.

9

Photon Interference

The Foundation of Quantum Communication

... it might be asked whether [two-photon] effects ... contradict the statement of Dirac that "each photon interferes only with itself. Interference between different photons never occurs." The answer is that they clearly do not ... each photon is to be considered as being partly in both beams ...

L. Mandel (1964)[1]

Controversy is good for physics—it's how conceptual progress is made on difficult topics. When two theories or two experiments seem to disagree, then deeper understanding is needed to resolve the paradox. The paradox of wave-particle duality in quantum physics, for instance, where quantum systems display particle properties in some experiments but wave properties in others, was resolved only after notions of partial information were folded into the interpretations of the phenomena. The more information you get (or could get in principle even though you may not actually look) about which path a particle has taken, then the less it behaves as a wave and the more it behaves as a particle, and vice versa. Another example is the controversy over Hanbury Brown–Twiss intensity interferometry discussed in Chapter 5. Its resolution by Purcell cautioned

Interference. David D. Nolte, Oxford University Press. © David D. Nolte (2023).
DOI: 10.1093/oso/9780192869760.003.0009

physicists to think more carefully about what photons are and what they do in optical systems, which helped to launch the field of quantum optics. However, some paradoxes in physics defy resolution, forcing physicists to take sides in continuing debates. The most famous of these is the paradox of Einstein, Podolsky, and Rosen—the EPR paradox. Proposed in 1935, with Einstein leading the charge against "God's dice,"[2] this paradox has yet to be resolved to everyone's satisfaction. Some say it is proof of parallel universes. Others say it must be caused by extra variables hidden inside quantum theory. Regardless, it has motived impressive interferometric experiments that challenge the very conception of reality. At the same time, it has provided powerful resources for the real-world applications of quantum communication and quantum computing.

Entanglement and the EPR Paradox

Despite the quantum hypothesis of Max Planck in 1900, Einstein can rightly be considered to be the father of the quantum when he introduced the quantum of light, what we call the photon, in his 1905 paper in *Annalen der Physik*.[3] Planck had understood the need for a quantum discontinuity to explain persistent problems in the theory of black body radiation, but it was Einstein who understood the fundamental need to quantize light into packets carrying $E = hf$ amounts of energy, where h is Planck's constant and f is the frequency of the light. Einstein was so dedicated to the physics of his photon that he even delayed his work on general relativity around 1909 to combat the hesitancy of other physicists to accept his arguments supporting the "reality" of the photon. Yet years later, as Bohr and Heisenberg and Born drove quantum physics forward, Einstein recoiled from the uncertainty and indeterminism that they progressively pushed. No revolutionary, he was still a student of the old mechanical universe of the nineteenth century, a mechanical universe of exact cause and effect without the need for rolls of the dice.

During a prolonged series of debates between Einstein and Bohr at conferences hosted by the industrialist Ernest Solvay through the 1910s and 1920s, Einstein sought to show that quantum mechanics was incomplete.[4] He attempted to argue that what looked like indeterminacy was just a lack of sufficient information—that there were variables hidden from the experimentalist that were nonetheless real, determining how each measurement would come out. However, Einstein was increasingly on the losing side of the arguments as his mechanistic universe was left behind. In 1935, he sought one last time to illuminate what he considered to be the most egregious flaw in Bohr's Copenhagen interpretation of quantum mechanics, a feature that today is called "nonlocality" while Einstein called it "spooky action at a distance."[5] This is where a quantum measurement at one location depends on a measurement at a different location, even if that other location is so far away that no effect of one on the other is possible.

The Copenhagen interpretation of quantum mechanics, though it can be stated in many different ways, is most simply defined through three principles:

1) Max Born's interpretation of the wavefunction whose squared modulus is the probability of observing an event;

2) Niels Bohr's wave-particle complementarity where wavelike experiments measure wavelike properties and particle-like experiments measure particle-like properties; and

3) John von Neumann's measurement principle in which an act of observation collapses a wavefunction to a specific state.

The last of these, wavefunction collapse, is where quantum wavefunctions, prior to a measurement, can exist in a superposition of states, but the act of measurement collapses all of the probability into only one of these states. An absurd part of wavefunction collapse, at least to Einstein, was the part where the collapse supposedly happens simultaneously, regardless of the "size" of the wavefunction. Although instantaneous wavefunction collapse carried no independent information, it still violated Einstein's common sense about the ultimate speed limit of physical processes set by the speed of light. Therefore, to explain the dependence of distant quantum measurements on each other, Einstein believed that quantum states must carry with them some set of hidden variables that would predict precisely the outcomes of experimental measurements, even if these variables were themselves inaccessible. Leading up to 1935, Einstein had generated progressively more ingenious thought experiments to try to thwart the new quantum mechanics. Now, in discussions with two colleagues, Boris Podolsky (1896–1966) and Nathan Rosen (1909–1995), at the Institute for Advanced Study, he constructed the ultimate challenge to Bohr.

Einstein, Podolsky, and Rosen (known collectively as EPR) proposed a paradox[6] in which wavefunction collapse seemingly violated Heisenberg's vaunted uncertainty principle. They were using one aspect of the Copenhagen interpretation to violate another, hence demonstrating its inconsistency. The ingenious part of the challenge—the part that was probably supplied by Einstein—is the use of two particles, sharing a single quantum wavefunction, that are so widely separated that no information can be transmitted between them during any measurement process. In Einstein's view, this meant that the measurement of one particle could have no causal effect on the other, hence the state of each particle exists *locally*, independent of whatever is happening to the other particle.

Einstein did not write the paper himself, which was left to Podolsky who buried Einstein's intuitive ideas underneath formal descriptions of measurements of position and momentum. For instance, by measuring the position or momentum of one of the particles, one could determine the associated property of the other particle to arbitrary precision through simple conservation laws.[7] Podolsky claimed that this violated Heisenberg's uncertainty principle. Bohr was able to defend his position in a paper only a few months later by pointing out that the EPR prescription required *either* position *or* momentum to be measured, but not both, and hence was fully in line with the uncertainty relation.

Einstein was a little miffed at Podolsky for losing sight of the essential point of locality and letting Bohr off the hook so easily.[8] He was so disappointed with the outcome of the EPR paper that he left the field of battle for good. Yet in its wake, the EPR paper established a fundamental quantum feature that Schrödinger later called *entanglement*,[9] where probabilistic measurements at two locations are dependent on each other, giving measurement outcomes that are correlated. An extreme example of entanglement is where one makes measurements on a large number of widely separated particle pairs, each measurement having a 50/50 outcome when viewed separately, and hence each set of measurements looking like the flipping of a coin, but when the two seemingly random sets of measurements are brought together, they have 100 percent correspondence to each other. Instead of two coins, there was only one! But even this description is too "classical," because if both sets of particles shared the same coin then that would be the same as Einstein's hidden variable. Quantum entanglement is more subtle than that, and it would stoke controversy and challenge theories of quantum reality and nonlocality for decades. Yet through its controversial nature, it became the central tool in experimental tests of quantum principles, especially after it was refined by a wayward physicist who showed with crystal clarity that quantum systems and classical systems could not be reconciled.

The American-British physicist David Bohm (1917–1992) was a deeply passionate physicist with a deeply troubled career. In 1951, as an assistant professor at Princeton, he was arrested and hauled before Senator McCarthy's Un-American Activities Committee for his communist sympathies. He avoided jail, but Princeton would not renew his contract. His former advisor from Berkeley, J. Robert Oppenheimer, who was the new director of the Institute for Advanced Study at Princeton, advised him to leave the country to let things settle down. But when Bohm landed in Brazil, the US consulate confiscated his passport. Marooned in a foreign country with no other recourse, he reluctantly swapped his American citizenship for Brazilian and obtained a faculty position at the University of Sao Paolo.

In the midst of this chaos in his personal life, Bohm was surprisingly productive. He completed an insightful textbook on quantum mechanics, building the whole edifice upon firm observational foundations, not introducing the Schrödinger equation until 150 pages into the book. As he considered the EPR challenge to Bohr, he retooled the thought experiment to replace the continuous observables of Podolsky with the discrete states of a spin-one-half system that could only be observed as spin up or spin down in a measuring apparatus (like the Stern–Gerlach effect). By restricting the possible outcomes of a measurement, Bohm was able to state the paradox in much simpler terms. The book was published shortly before Bohm was forced to flee to Brazil, and Einstein was so pleased with Bohm's open-minded view of quantum interpretations, that he invited him to his office to talk it over.[10] This discussion profoundly affected Bohm as he better understood Einstein's distaste for instantaneous wavefunction collapse.

Bohm began developing an alternative quantum theory to stand against the problems of the Copenhagen interpretation. His theory was a wave-particle description of quantum mechanics that contained Einstein's hidden information and removed the fundamental uncertainty of God's dice. It was clear to Bohm that Einstein's hidden variables could be contained in a quantum potential that extended across space in a *nonlocal* hidden variable theory. This was a compromise that still rejected Einstein's intuition about locality, but it removed the indeterminism that he deplored. Bohm's theory relies on the Schrödinger equation for the wavefunction, just as in the Copenhagen theory, but in Bohm's theory the wavefunction defines a quantum potential that drives the dynamical trajectories of individual particles. There is no indeterminacy, because each initial condition launches a unique particle trajectory through the space. There is no wavefunction collapse because each particle is measured somewhere on its trajectory. What looks like indeterminacy is just sensitivity to initial conditions (SIC), which is a concept familiar to us today from the physics of deterministic chaos. The slightest change in an initial condition causes the particle to take a radically different trajectory. To fully understand the character of a system, many particles with many initial conditions must be sent through the system in an ensemble approach characteristic of mainstream quantum theory. Bohmian mechanics predicts the same things as Copenhagen mechanics, but it is done without wavefunction collapse.

Bohm published two papers on the new formalism[11] shortly after he was established in Brazil in 1952. A few months later he received a letter from de Broglie. It turned out that Bohm's new approach had similarities to an old idea put forward by de Broglie at the 1927 Solvay Congress that he had called a "pilot wave." The pilot wave was similar to Schrödinger's wavefunction, but rather than using Max Born's probabilistic interpretation of the wavefunction, de Broglie proposed that the pilot wave directed the motion of particle trajectories. Unfortunately, this was at the same time when Heisenberg was denying any existence to any trajectory, and the gang from Copenhagen could not accept de Broglie's proposal. Under their onslaught, de Broglie retreated and never pursued the mathematical consequences of his pilot wave.

Although Bohm was supported enthusiastically by his Brazilian colleagues, he felt isolated, and he immigrated to Israel in 1955 to work at Technion University in the verdant Mediterranean port town of Haifa. There he met Yakir Ahronov (1932–), and the two began an extraordinary collaboration. One of the gems that came from their association was the Ahronov–Bohm effect of the magnetic vector potential.[12] This bold theory predicted that the wavefunctions of quantum particles, that experienced no magnetic field and hence no physical forces, would nonetheless be influenced by changes in a magnetic field *somewhere else* away from the particles. This physical effect looks nonlocal, unless the magnetic vector potential is accepted as something real and more than a mathematical construction. In the Ahronov–Bohm theory, the magnetic vector potential ends up looking somewhat like a hidden variable.

During his time in Israel, Bohm continued to think about his quantum potential and other possible hidden-variable theories. Working together, Bohm and Ahronov found an even simpler example of an EPR system that involved the emission of two photons that could be analyzed with optical polarizers[13] (see Figure. 9.1). As they looked into it, they realized that the famous

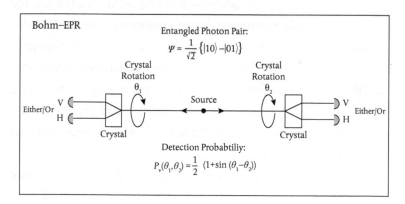

Figure 9.1 Bohm–Aharonov version of the EPR system. A source generates a two-photon entangled pair that separate so far apart that there can be no causal effect of one on the other. They are analyzed through optical crystals (like feldspar) that split vertical and horizontal polarizations. The experimental outcome behaves "as if" the measurement of one photon instantaneously polarizes the other.

Chien-Shiung Wu, who later became known as *Madame Wu*, had performed such an experiment in 1950 on positronium decay into two gamma photons.[14] Bohm and Ahronov reanalyzed her experimental results in terms of the EPR paradox, and they showed that the data agreed with the nonlocal predictions of quantum theory. Most importantly, they showed that the pair of photons carried the strange correlations of entanglement.

Shortly after their EPR paper in 1957 and shortly before they discovered the Ahronov–Bohm effect in 1959, Bohm left Israel for a faculty position at the University of Bristol in the UK and then to the University of London where he finished out his career. This happy landing for the American ex-pat never fully erased his ignominious treatment by the American government, nor fully restored him in the eyes of the physics community. He remained a philosophical gadfly, tilting at the Copenhagen school as well as bigger targets, such as the purpose and responsibilities of science in society. His hidden variable theory continues to be the foundation of "Bohmian Mechanics" today, but it is a decidedly minority view of quantum theory. His most lasting impact, simple as it was, was his crystal clear formulation of the entanglement problem, replacing Podolsky's convoluted arguments with a Stern–Gerlach analogy that confronts every student of physics who seeks to understand nonlocality.

So far in this chapter, descriptions of entanglement have only been qualitative, referring obliquely to "correlations" and "dependences" that may seem vague. However, quantum entanglement is surprisingly easy to define mathematically, especially when there are only two particles (either A or B) and each can be in only one of two possible states (either a 0 or a 1). In this case, entanglement is simply a statement of "either/or." Either particle A is in state 0 and particle B is in state 1, or particle A is in state 1 and particle B is in state 0. Expressed in symbolic wavefunction form, one example of an entangled state is

$$\Psi = \frac{|0\rangle_A |1\rangle_B - |1\rangle_A |0\rangle_B}{\sqrt{2}}$$

This uses Dirac's "bra-ket" notation, where a specific state, such as state 1, is enclosed in the "ket" $|1\rangle_A$ and the subscript A refers to one of the two particles. (The square-root of 2 just normalizes the associated probabilities to unity.) If we pick a convention such that the A state always precedes the B state, then this expression becomes even simpler as

$$\Psi = \frac{|01\rangle - |10\rangle}{\sqrt{2}}$$

In contrast, a *separable state* may be written (keeping the subscripts) as

$$\Psi = \frac{|0\rangle_A |0\rangle_B - |0\rangle_A |1\rangle_B}{\sqrt{2}}$$
$$= |0\rangle_A \frac{(|0\rangle_B - |1\rangle_B)}{\sqrt{2}}$$
$$= \Psi_A \cdot \Psi_B$$

In this case, the wavefunction is the product of a distinct state for particle A, and a separately distinct state for particle B, allowing the state or wavefunction of each particle to be measured independently from the other. The entangled state, in contrast, cannot be factorized or expressed as a single product. This is what makes the entangled state so special. Both the entangled and separable wavefunctions are quantum superpositions, but in the entangled state there is no single "reality" for either particle. Each particle's state is conditional on the other state. Neither particle has a distinct state prior to some measurement that condenses, or collapses, the individual particles into distinct states.

If we think of two hypothetical experimentalists—call them Alice and Bob—who operate their own polarizers and detectors on their own half of the particle pair in the experiment of Fig. 9.1, then in the spirit of the Copenhagen tradition we can say that when Alice gets a result she instantly defines the probabilities for Bob to get state 0 or 1. Equivalently, we can say that when Bob gets a result, he instantly defines the probabilities for Alice to get state 1 or 0. The particles can be separated as far apart as needed, and in Copenhagen "lingo" one could say that the measurement of one particle instantaneously collapses the state of the other. Despite the arbitrarily large "speed" of this effect, what Einstein called "spooky action at a distance,"[15] it is easy to show that no information can be transmitted in the process, softening the blow to local realists. Even so, if the notion of instantaneous collapse across arbitrary distances is still unpalatable, then it may be safer to say that the measurement probabilities of Alice and Bob are correlated, regardless of who makes the measurement first, and hence there is no need to think of one collapsing the wavefunction of the other. The correlations are the "reality." The correlations result "as if" one measurement collapses the other state and vice versa—but the correlations are what is primal about the quantum pair, not the collapse. However, for those who found *both* instantaneous collapse *and* primal correlations unpalatable, they may have hoped for a hidden variable to explain everything, but they hoped in vain, because an Irish physicist would definitively abolish all such hopes.

John Bell's Inequalities

John Stewart Bell (1928–1990) was born to working-class parents in a Protestant neighborhood of Belfast, Ireland.[16] In the days before education reform in the UK, he was the only one of four siblings to receive an education beyond the age of 14. But John loved school and declared at the age of 11 that he wished to be a scientist. His family sometimes called him "Prof" at home because of the endless tidbits of arcane knowledge that he possessed. Although his parents could not afford to send him to secondary school, he won a scholarship to attend a vocational school where he took trade classes, including one on brick laying. He excelled in the courses needed to apply for university but, lacking the funds to attend after his graduation in 1944, he took a job as a laboratory technician in the physics department of Queens University in Belfast. Once again, he stood out, and the physics professors running the lab helped him obtain a grant to attend college.

One of these was the junior professor (a Reader) Richard Sloane from whom Bell took many of his physics classes, including quantum mechanics. In the late 1940s, quantum theory had become dogmatized into the orthodox view of the Copenhagen interpretation, but it still lacked a precise mathematical description for both Bohr's complementarity principle and von Neumann's wavefunction collapse. At that time, physicists talked of the "Heisenberg Cut" which was a dividing line between the quantum systems being studied and the classical measuring apparatuses that were used to study them. Sloane was typical of the day, able to parrot Bohr and Heisenberg, while lacking a deep intellectual grasp of the inconsistencies lurking in the problems of quantum measurement. The ironclad hold that Bohr's ideas had on

quantum interpretation was best summed up by Paul Ehrenfest who witnessed the Bohr–Einstein debates. He once wrote,
"Brussels-Solvay was fine! ... BOHR towering over everybody.
At first not understood at all ..., then step by step defeating
everybody. Naturally, once again the awful Bohr incantation
terminology. Impossible for anyone else to summarize."[17] Bohr
won his arguments by sheer weight of words rather than precise
mathematical arguments, and this had become the dogma.

Bell, even as an undergraduate physics student, felt that the
dogma of the day swept important details under the rug. When
he confronted his teacher and long-time benefactor on the meaning of the Heisenberg uncertainty relations, they got into a heated
argument. In an interview with the science writer Jeremy Bernstein,[18] Bell recounted how Sloane spouted the party line as Bell
kept attacking it, until they were nearly shouting at each other.
Bell remembered, possibly with some regret, that he had gone so
far as to accuse Sloane of intellectual dishonesty by his refusing
to look behind the curtain of Bohr's "awful incantation terminology." This story is prophetic, because it illustrates how deeply
Bell already was thinking about the foundations of quantum
theory at a time when no-one else was doing so. Bell believed
with Einstein that there must be hidden variables to account for
apparent uncertainty, despite an argument by von Neumann[19]
that was supposed to have proven that *all* hidden variable theories
were in conflict with the predictions of quantum theory.

After Bell graduated from Queen's University in 1948 with dual
degrees in experimental physics and mathematical physics, he
could not afford to go to graduate school, so he took a position
with the Scientific Civil Service where he was assigned to work on
problems of accelerator physics. He was soon considered a genius

at solving accelerator design problems, but he still harbored an interest in the foundations of quantum physics. Bell was delighted when Bohm published his hidden-variable theory paper in 1952, creating something that von Neumann had said was impossible, and it reignited Bell's own interest in hidden variables, which became a hobby he pursued in his free time away from work.

Because of his excellent performance on accelerators, Bell was granted a leave of absence to attend graduate school at Birmingham, where he studied under Rudolph Peierls. When asked what topic he wished to pursue for his PhD, Bell suggested the foundations of quantum physics, but Peierls steered him away. No-one was studying the foundations of quantum physics at that time, and to do so would guarantee unemployability. Bohr and Copenhagen were too deeply entrenched, and to assail them would be like tilting at windmills. (Look what happened to the exiled Bohm!) Therefore, Bell chose quantum field theory instead. He became interested in time-reversal symmetry, which was closely associated with two other symmetries of fundamental particles known as charge conjugation and parity. Collectively they are called CPT symmetry, and Bell showed that CPT symmetry was a fundamental symmetry of physics that must be obeyed by all particle interactions. The CPT theorem in whole requires that a violation of time-reversal symmetry (T) guaranteed a violation of CP. This violation has indeed been observed in the decay of neutral kaons and is thought to be partially responsible for the composition of our universe being exclusively of matter instead of equal amounts of matter and antimatter. Bell's work in this area would have been groundbreaking if Wolfgang Pauli had not gotten to the same results about one year earlier. Bell received his PhD in 1956, and in 1960 he moved to the European high-energy facility at CERN in Geneva, Switzerland, to work on the theory of particle physics.

The pivotal opportunity in Bell's career came during a sabbatical to the Stanford linear accelerator (SLAC) from 1963–1964 when he finally had free time to spend on his "hobby" of hidden variables. He first showed that von Neumann's "proof" of the impossibility of hidden variables contained a mathematical error that not only invalidated the proof but opened the door to nonlocal hidden variable theories like Bohm's. Bell then derived a series of inequalities that must be obeyed by any local realist hidden variable theory. The genius of these derivations is both their simplicity and their generality—an outline of the inequality is shown using Venn diagrams in Figure. 9.2. The inequalities were derived without needing to know anything about the details

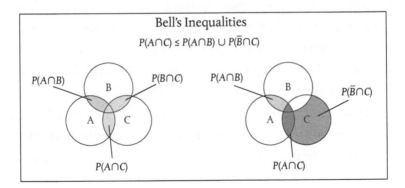

Figure 9.2 Bell's inequalities for classical probabilities described as a Venn diagram of independent probabilities. The diagram on the left shows three possibilities and their overlaps. On the right, the chances of A and C both being true must be less than the chances of A and B or Not-B and C being true. This common-sense diagram is violated by quantum systems because quantum probabilities are not independent but depend conditionally on each other.

of the actual hidden variables, allowing entire classes of theories to be considered at a time. Bell published his results in an obscure journal[20] and returned to his day job at CERN calculating particle interactions.

At first no one noticed, but five years later—and out of the blue—he received a letter from the American physicist John Clauser (1942–) who was then a graduate student in astrophysics at Columbia University in New York. Clauser proposed an experimental test of Bell's inequalities that would use atoms and optics, and he wanted to know whether Bell believed the experiment had merit. Bell was happy to reply, and he expressed interest in the experiment, not realizing that this first step would ultimately launch him and his theory to fame while starting a revolution in physics that would evolve into the field of quantum information science.

John Clauser came from a long line of Cal Tech scientists. His father and uncle were both scientists at Cal Tech, working with von Karmen on nonlinear fluid instabilities, and John attended Cal Tech as an undergraduate physics major. When he got to his quantum mechanics classes, he had deep discussions with his father who was suspicious of the standard interpretations of quantum theory, instilling in him a healthy skepticism toward the subject. This skepticism may have been one reason why Clauser had to repeat his graduate-level quantum course at Columbia University three times before getting it right.[21] During that graduate course, he was driven to look deeper into the subject as he struggled to master it, and he stumbled on Bell's obscure paper in the obscure journal and grasped its importance. Clauser began working out for himself how experimental systems might be able to test the theory. Just like Bell, Clauser was supposed to be working on other things, like his PhD dissertation in astrophysics. Therefore, when his advisor Patrick Thaddeus realized that Clauser's research was flagging, Thaddeus warned him to give up the quantum stuff—it was a dead end. Everyone knew that quantum mechanics worked, so there was no point testing it yet again. But Clauser had taken hold of the tiger's tail, and like Bell before him, it became his consuming hobby.

At that time Madame Wu, who had performed the positron annihilation experiments in 1950 that were analyzed by Bohm and Ahronov in 1957, was at Columbia, and Clauser dropped by her office, asking whether she had unpublished data on gamma polarizations. She did not, but she sent Clauser to talk with one of her graduate students who was repeating the experiment. During the discussion it became clear that the polarization detection efficiencies in high-energy physics were too low to reliably test Bell's inequalities, so Clauser began thinking in terms of optical experiments that could be controlled much better. He took a train up to MIT to talk with the quantum optics specialists Dan Kleppner and David Pritchard. By coincidence, one of Pritchard's new post-docs was Carl Kocher who had done an experiment at Berkeley for his PhD under Eugene Commins. The experiment had interrogated the orthogonal polarizations of a two-photon cascade from calcium atoms as an optical analog to Madame Wu's high-energy experiment. They had found the same results as Wu and had left it at that—there seemed to be nothing more to look at. Clauser knew better! Now he knew exactly how to modify the Kocher–Commins experiment to test Bell's inequalities. This is when he sent his letter to Bell, and other letters to both Bohm and to de Broglie. When he got positive reactions from them all, he wrote up his proposal as an abstract for the meeting of the American Physical Society (APS)—and a few weeks later received a phone call from Abner Shimony at Boston University.

Abner Shimony (1928–2015) was a quantum philosopher who spent as much time teaching philosophy in the philosophy department as physics in the physics department. Philosophers who delve into the foundations of quantum theory are tolerated a bit more than experimental physicists. Shimony also had become aware of Bell's paper, and his graduate student Michael Horne was working on the same problem of practical tests of the inequalities. Shimony and Horne had felt that their work was not quite ready for an APS abstract and had not submitted one, but when they saw Clauser's abstract, they knew they were being scooped. Rather than competing, Shimony decided to suggest a collaboration, which was the purpose of his phone call to Clauser. Shimony's ace up his sleeve was his connection to an experimental group at Harvard and the graduate student Richard Holt who was adapting the same Kocher–Commins apparatus to test Bell's theorem. Clauser, who had no access to such experiments, knew it was better to share the credit in a collaboration than risk being left behind, so he agreed and took the train to Boston for discussions. The collaboration now was composed of four people: Clauser, Horne, Shimony, and Holt who later became known collectively through the acronym CHSH. The essential new work that came out of the four-way collaboration was an experimental methodology that would get around the inefficiencies of photon detection to provide high-accuracy and high-confidence tests of the Bell inequalities. With the details worked out in the face-to-face meetings, it was left to Clauser to write it up for publication.

At that time, Clauser was living on a sail boat moored in New York City's East River. After he defended his thesis in astrophysics at Columbia and was offered a post-doc position at Lawrence Berkeley Laboratory in Berkeley, Clauser decided to sail there. The plan was to sail around Florida to Galveston, ship the boat to LA, and continue up the Pacific Coast to Berkeley. As he worked his way down the Atlantic, he stopped in ports-of-call that had been previously agreed upon, picking up comments and feedback sent to him from the CHSH team, and mailing drafts back.[22] The paper was nearly written when he ran into Hurricane Camille out in the Atlantic in mid-August 1969 and had to seek shelter in Florida. He abandoned the trip and shipped the boat from Florida, submitting the paper when he finally showed up at Berkeley.

Despite the head start that Holt had on the experiments at Harvard, once Clauser was in Berkeley, he had direct access to Commins himself and suggested doing the experiment. At first, Commins was not interested, and at one point said, "What a pointless waste of time all of that was."[23] But Clauser was shrewd enough to enlist Charles Townes, the Nobel Laureate of laser fame, to his cause. After a meeting with Commins, Townes put his arm around his shoulder and said, "Well, what do you think of this, Gene? It looks like a very interesting experiment to me."[24] And that was that. If Townes thought something was interesting, then it was hard to say no. Commins relented and loaned equipment, along with his graduate student, Stuart Freedman, to what was hoped would be a quick and easy demonstration so that everyone could get back to what they were actually paid for. But experiments never go that way, and the test was not completed until 1971. Fortunately, they beat the Harvard group despite the delays, and the Freedman and Clauser paper was published in 1972 as the first experimental test of the Bell inequalities.[25] In it, they reported experimental results that violated Bell's inequality by five standard deviations—in other words, they got precisely what quantum mechanics predicted. Tests by other groups followed, confirming and improving upon the Freedman and Clauser experiment with increasing accuracy, especially a series of experiments by Alain Aspect in Paris that increased the confidence to more than 10 standard deviations.[26]

This was no news to most physicists, who shrugged. In the 50-year history of quantum physics, every time anyone had gone up against the Copenhagen interpretation, they had either lost or had come up with nothing better. If the tests of Bell's inequalities were meant to find something new, then they failed and would have been relegated to the arcane tomes of quantum philosophy and received no further notice by physicists. However, these tests of the Bell conditions did more than that. First, as the optical techniques used to test the Bell inequalities became progressively more refined and efficient, they became the backbone of the quantum optics research techniques used today. Second, they drove home to physicists what everyone had known but had not been able to get the heads around: that quantum physics is intrinsically nonlocal. This means that it violates what might be called "common sense" and hence can do bizarre and unintuitive things. Rather than being merely novelties, these bizarre and unintuitive things became tools of quantum optics that can do things that classical systems cannot. This was especially true in the creative hands of a refugee of Nazi Germany who bent these tools to the service of nonlocal interferometry.

Leonard Mandel's Two-Photon Magic

Leonard Mandel (1927–2001) was born in Berlin at the end of the roaring twenties as Germany plunged into financial depression and a mean anti-Semitic national socialism gripped the country. In the late thirties the family fled Nazi Germany, first to London and then to Llangollen in Wales where Mandel was exposed in school for the first time to physics—a subject in which he excelled. Returning to London later in the war, but not so late as to avoid nights hiding in the Underground during V2 attacks, he attended a private school where he rose to first boy. His academics were so good that all of his teachers assumed he would receive a scholarship to attend Cambridge, but it never came. Instead, he worked nights and attended classes during the day at the University of London. It took him only two years to graduate with a dual bachelor's degree in mathematics and physics and another three years for a PhD in cosmic ray physics. After a few years in industry, he landed a lectureship in 1955 at Imperial College in London where he met Emil Wolf (1922–2018). Wolf was a fellow war refugee who had worked with Max Born in Edinburgh, Scotland, publishing a book together that is considered to be the bible of classical optics and coherence, referred to widely simply as *Born and Wolf*.[27] Mandel and Wolf began joint research on optical coherence, and when Wolf moved to the University of Rochester in the US in 1959, Mandel followed him five years later, arriving the same year that Bell published his inequality paper. Mandel was receptive of Bell's arguments, because he had already been thinking of coherent aspects of quantum optics after the experiment of Hanbury Brown and Twiss had caught his eye some years earlier.

Optical coherence is a strange and subtle subject. The coherence of a laser beam is what gives it its high intensity and beam-like qualities, making it strikingly different than the diffuse light of an incandescent light bulb. But even incandescent light has some residual coherence. After all, Young performed his double-slit experiment with sunlight, which is a purely thermal source and might be thought to have no coherence at all. Yet with a judicial use of pinholes and filters, one can coax residual coherence out of even sunlight to create interferometers. In this sense, one might say that there is no such thing as incoherent light. Mandel extended these ideas to the single-photon limit of light, developing a deeply intuitive understanding of quantum coherence, and when the laser was invented in 1960 by Ted Maiman at Hughes Research Lab in Malibu, Mandel and his group at Rochester were among the first to use it to explore the coherence of laser light.

In 1967 Mandel and his student R. L. Pfleegor shocked the optics world with an experiment that observed interference between two independent lasers. In the experiment, they directed light from *two* HeNe lasers to cross at a small angle.[28] This is in contrast to splitting a *single* laser beam with a beam splitter and crossing the two beams at small angles that produces a dramatic series of bright and dark intensity fringes just like Young's fringes. That interference is a simple and direct consequence of the mutual coherence of two beams from the *same* laser. But two *different* lasers should have no shared coherence, and therefore, when their beams cross, the belief had been that there would be no intensity fringes.

Nonetheless, Mandel and Pfleegor were able to tease out interference through a couple of clever tricks associated with the stability of the lasers and with the operation of the photomultiplier tubes. They attenuated the photon flux to such a low level that one photon would be absorbed by a PMT before a second photon had even entered the apparatus. This brought their experiment into the extreme single-photon limit. The biggest surprise was that, despite there being no "overlap" of the photon arrivals, they *did* observe an interference pattern. This seemed to violate Dirac's old adage that "… each photon interferes only with itself. Interference between different photons never occurs,"[29] launching a small controversy. Mandel resolved this by pointing out that *pairs* of photons can interfere with themselves in the sense that a two-photon wavefunction can be a superposition of indistinguishable paths to the detector. When a PMT fires, it is not always possible to know from which laser the photon originated, and there is an interference between the possibilities. This new insight launched an active subfield of quantum optics focused on multi-particle coincidence effects, a research area facilitated by the invention of a unique new light source called parametric down-conversion.

Parametric down-conversion is a nonlinear-optical effect in which a photon enters a transparent non-centrosymmetric crystal and splits into two lower-energy photons. The process conserves both energy and momentum, causing the generated photons to leave the crystal at different angles, making it easy to use apertures to create two spatially separated beams of correlated photon pairs. For archaic reasons, the two beams are called the signal and the idler. In the limit of low pump power, down-conversion generates single correlated pairs of signal and idler photons that can be used as a guaranteed source of photon pairs in downstream experiments.[30] In the mid 1980s, parametric down-conversion provided a radical shift in quantum optics research.

Mandel had an uncanny sense of higher-order coherence effects in the physics of light, and with his student Rupamanjari Ghosh in 1986, he demonstrated non-classical two-photon interference effects.[31] The interference was observed in coincidence detection of the two photons even when there were no visible interference effects in the single-photon intensity alone. Once again this was a shock to the optics community whose mindset was largely informed by classical coherence. The experimental configuration is shown in Figure. 9.3 in idealized form. The signal and idler beams are directed to overlap in a region where the beams cross. If a single detector is scanned across the region, the intensity it measures is constant—without interference.[32] However, if *two* detectors are scanned across the overlap region and fed into a coincidence counter, then the *coincidence rate* shows a sinusoidal

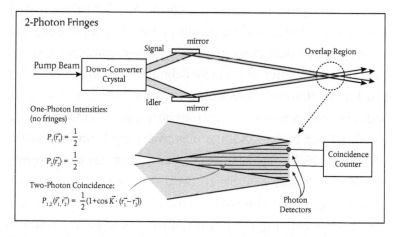

Figure 9.3 Experimental configuration of Ghosh and Mandel. The signal and idler beams from a down-conversion crystal overlap in a detection region. The measured intensity is constant across the region, but the two-photon coincidence rates vary sinusoidally in a two-photon fringe pattern.

fringe pattern as a function of the separation between the detectors. This effect displays intensity interference, like the HBT experiment, but *unlike* HBT using a classical light source, with down-conversion there is no associated amplitude interference! The two-detector coincidences in the Ghosh–Mandel experiment map out high and low spatial fringes of coincidences *that are invisible to a single detector*. This experiment shifted how quantum coherences were understood.

The Ghosh–Mandel two-photon interference experiment using the down-conversion crystal immediately suggested a new type of two-photon interferometer. The demonstration was done in Mandel's group a year later in 1987. The authors of the paper were Chung Ki Hong, Zhe-Yu (Jeff) Ou, and Leonard Mandel, and the experiment and the specific configuration are now routinely called the HOM interferometer. The experimental arrangement is shown in Figure. 9.4 where the signal and idler beams from a down-conversion crystal are incident on the two input ports of a beam splitter, which plays the role of a beam combiner (like the second beam splitter of a Mach–Zehnder interferometer). With two input photons, there would be three output possibilities: both photons exiting together from port 1, both photons exiting together from port 2, and one photon exiting from each of the ports (for which there are two alternate ways this can happen). In addition, it was possible to delay the entry of one photon relative to the entry time of the other. When the delay between the photons entering the beam splitter was large, the photons would exit separate ports randomly just as expected classically. However, when the two photons were incident simultaneously, and hence overlapped in both coherence and coincidence, there was destructive interference between the two different ways the

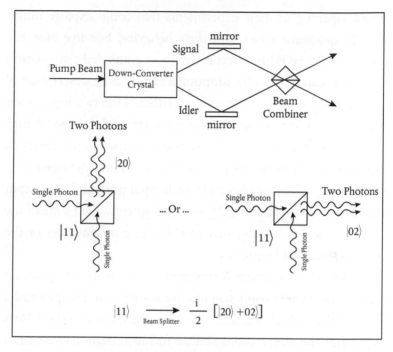

Figure 9.4 The HOM interferometer. A two-photon input on a beam splitter generates an entangled superposition of the two photons exiting the beam splitter always together.

photons can exit the two different ports, and HOM observed that both photons always left the beam splitter together in pairs from the one port or the other but never both. In other words, the coincidence counts for the two PMTs, one at each output side of the beam splitter, vanished when the two photons entered the beam splitter at the same time. The photons bunch up just as bosons tend to do, but now the bunching is done deterministically one pair at a time.

The ability to prepare quantum input states with known numbers of photons represented a valuable new tool for quantum optics, opening up new experiments that could explore multi-particle quantum states and their behavior. For instance, the output of the HOM interferometer is an entangled two-photon state. This can be seen by adopting a *number mode* description of the quantum state. A mode refers to either of the two input ports of a beam splitter, or the two output ports, and the modes may be occupied by zero, one or two photons in the case of the HOM experiment. In this number mode representation, the input state is $\Psi_{in} = |11\rangle$ with one photon in each input port, and the output state is $\Psi_{out} = \frac{1}{\sqrt{2}} [|20\rangle + |02\rangle]$ which is an entangled state of the two photons leaving together in either one output port or the other as shown in Figure. 9.5.

The HOM experiment introduced a new form of quantum interferometry—two-photon interferometry—and it spawned a flood of theoretical predictions[33] and experimental tests.[34] Next steps after the single beam splitter included demonstrations of Michelson and Mach–Zender interferometers. In addition, there was a broad range of signal and idler beam configurations that could be used to generate entangled two-photon states that could be selected by polarization, by energy or by direction. These experiments led to increasingly sophisticated tests of quantum coherence.

A particularly dramatic use of two-photon interferometry reconnected with the intrinsic nonlocality of quantum physics that had been pursued by John Clauser in the early 1970s. An experiment was performed by Mandel's graduate student Jeff Ou[35] in 1989 using a double Michelson interferometer configuration following a suggestion by J. D. Franson at Johns Hopkins.[36] The idler beam enters one Michelson interferometer, and the signal beam enters an entirely independent Michelson interferometer,

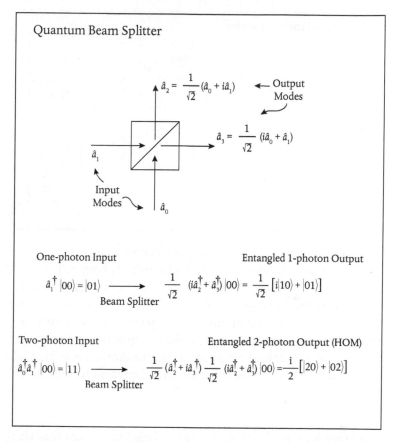

Figure 9.5 Quantum operations of a beam splitter. A beam splitter creates a quantum superposition of the input modes. The $\hat{a}^\dagger$ and $\hat{a}$ symbols are quantum number operators that create and annihilate photons. The symbol $|00\rangle$ is the vacuum state. A single-photon input produces an output that is a quantum superposition of the photon coming out of one output or the other. Two photons in each input mode produces a two-photon entangled output known as the HOM output.

shown in Fig. 9.6. A key feature of the experiment is that each interferometer is "unbalanced," so that the delays between the two arms of each Michelson interferometer are much larger than the

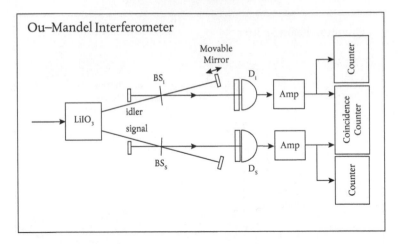

Figure 9.6 The Ou–Mandel interferometer consisting of two Michelson interferometers for an entangled pair of photons. The interferometers are each unbalanced with a long arm and a short arm and a path difference larger than the pulse duration, hence no intensity interference is observed at the individual detectors. However, there is strong interference in the coincidence detections between the two detectors, because either both photons take the short path, or both take the long path, arriving at the detectors at the same time for either case.

coherence length of the light source. This guarantees that there are no field-interference effects, which is tested by monitoring the individual count rates at each of the detectors. The misbalance is adjusted to be the same for both interferometers, although one of the mirrors can be shifted for path-length scanning, and as the mirror is displaced, the intensity at the detectors remain steady without any interference fringes.

The crucial part of this two-photon and two-interferometer experiment was the coincidence rate from the two detectors. Each single-detector firing rate remained steady as the mirror was shifted, while the two-detector coincidence rate changed sinusoidally with displacement. As with the Ghosh and Mandel experiment, this goes beyond the usual HBT intensity interferometer with classical light sources which relies on intensity correlations that are connected to field-based interference effects. Both field-based and intensity-based interference effects exist side by side for a classical light source. But in the Ou–Mandel experiment, these interference effects are complimentary: when field-based interference is observed, then the intensity-based effects vanish, and conversely when the intensity-based interference is observed then the field-based effects vanish.

There is more subtle behavior in the Ou–Mandel experiment that connects to the nonlocality of quantum physics. Consider when both Michelson interferometers are adjusted to have the same imbalanced arms, each with a long arm and a short arm, and the coincidence rate on the two detectors is a maximum. This means that if a photon exits the beam splitter toward a detector in one of the Michelson interferometers, the second photon also exits toward the detector in the other Michelson. Conversely, if a photon exits the beam splitter back toward the source in one Michelson, then the second photon does the same in the other. It is "as if" each photon knows what the other is doing and follows suit even though the two Michelson interferometers are far enough apart that there can be no causal influence of one on the other. How do the photons know? Obviously, they don't, but they are correlated through entanglement and the nonlocality of the EPR effect! One could say, in Copenhagen parlance, that measuring one of the particles collapses the *choice of path* of the other. Or one could say, in more neutral language, that the paths themselves share nonlocal EPR correlations. Either way, this and other multi-photon coherence experiments from the Mandel group rewrote the book on the physics of optical coherence—literally! In 1995 Mandel and Wolf published the quantum optics successor[37]—*Mandel and Wolf*—to the bible of classical optics—*Born and Wolf*—that had been published in 1969. Mandel's work set up a brave new world of quantum optics that led ultimately to quantum communications and quantum computing, beginning with a technique that can teleport coherence across space and time.

"Beam me up Scotty"

In 1964 Gene Rodenberry, a successful but combative Hollywood screen writer with outrageously bushy sideburns, pitched a space-western TV series that he called *Star Trek* to Desilu Productions. Desilu was founded many years earlier by Desi Arnez and his wife Lucille Ball in their Vaudeville days when Arnez was the headliner and Ball was his unknown sidekick. By the mid-60s Arnez and Ball were divorced, Lucile Ball was one of the biggest names in television, and Desilu had become one of the most successful production companies in Hollywood. Space, in the middle of the space race of the Cold War, was a hot item, and Desilu picked up Rodenberry's *Star Trek* and pitched it to NBC, who paid to film a pilot "The Cage." The pilot failed and never made it to TV in its original form, but the NBC executives thought the idea had promise and paid for a second pilot "Where No Man Has Gone Before." This pilot had more energy and a stronger cast, starring William

Shatner as the captain, and sleeker sci-fi backdrops, helping it to land a weekly spot in prime time. *Star Trek* aired for three seasons and 79 episodes, the last episode airing on June 3, 1969.

During production of the first pilot, getting the crew down to the planet and up again by shuttle craft took up too much story time and required expensive special effects, so Rodenberry "invented" the transporter as a quick and cheap expedient. With the success of the series, the transporter became a cultural fixture best remembered through the catch phrase, "Beam me up Scotty," that everyone thinks was said by captain James Kirk to his chief engineer, even though those exact words were never spoken in the original series nor in the subsequent movies. The transporter became part of the science-fiction milieu, sparking the imagination of a generation of budding scientists who might dream of turning fiction into reality. Ten years later in 1979, on the heels of a wildly successful debut of George Lucas' *Star Wars*, Rodenberry released a movie reboot of his cancelled series featuring most of the original cast.

That same summer, as moviegoers watched Captain Kirk and his crew teleporting to and fro, Gilles Brassard (1955–), a Canadian computer scientist from Montreal, was floating motionless in the ocean off a beach in Puerto Rico when a complete stranger swam up to him and started telling him wild ideas about quantum information. Brassard listened at first politely, wondering who this kook was, but was slowly drawn in by the fascinating ideas the stranger was telling him.[38] By the time they had both swum back to the beach, they had invented the world's first quantum cryptography protocol. It became known as BB84—one of the Bs was Brassard, and the other B was the stranger in the water, Charles Bennett (1943–), who was working at the famous Thomas

J. Watson Research Center of IBM nestled in the wooded enclave of Westchester County, north of New York City.

At the time of their encounter, Bennett and Brassard were both attending a conference on the future of computing, held that summer in San Juan, Puerto Rico. Bennett was a physicist studying the limits of computing, and he was beginning to explore quantum ideas related to work of Stephen Wiesner (1942–2021), a classmate of his at Brandeis University, who had invented a way to make a quantum currency that could not be counterfeited. Bennett thought there should be a way to use Wiesner's ideas to send information over communication channels without the risk of an eavesdropper copying the information. Bennett knew who Brassard was, and had seen him at a reception, but had not approached him at that time. It was later, on the beach when he saw Brassard out in the ocean that he decided to take a chance to pick his brain. As a physicist, although he was somewhat unfamiliar with techniques in computer science, he knew he would need help to dig deeper if he were to develop his ideas on quantum communication. Brassard, it turned out, was exactly the right person to approach, if in such an unexpected way, and together they solved the problem. It took Bennett five years to build a test system at IBM, which he demonstrated for the first time in 1984—hence the protocol's name of BB84.[39] Other, improved, protocols for quantum cryptography came steadily over the following years as Bennett refined the problems of quantum communication.

The operation of the BB84 protocol is illustrated in Figure. 9.7 using optical polarization as the encodings. There are two polarization bases: horizontal (H) and vertical (V) for 0 and 1 encoded in the "+" basis, and the two diagonals (D+ and D–)

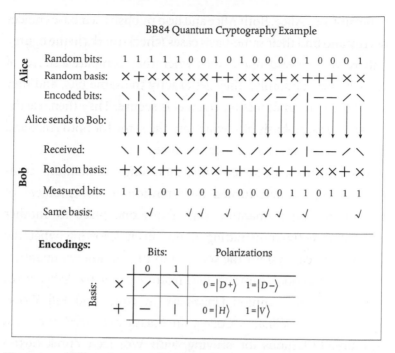

Figure 9.7 BB84 quantum cryptography protocol. Alice randomly chooses bits and the bases to send polarized photons to Bob. Bob chooses a random basis and measures the bits. They publish their bases openly and find their same-basis cases (check-marks). If there is no eavesdropper, then when they compare a subset of bits (chosen randomly), they get exactly the same bits, and they know their communication channel is secure. They then use the remaining same-basis bits as an encryption key that they both share.

for 0 and 1 encoded in the "×" basis. The communication takes place between two people who are traditionally called "Alice" and "Bob" in cryptography circles. Alice randomly chooses a string of bits, 0s and 1s, and also chooses an equal number of polarization bases, + and ×, to send polarized photons to Bob. Bob likewise chooses random polarization bases and measures the arriving photons from Alice. Both Alice and Bob publish their base choices openly and find their same-basis cases (check-marks in the figure). If there is no eavesdropper, then when they compare a subset of bits (chosen randomly), they get exactly the same bits, and they know their communication channel is secure. They then use the remaining same-basis bits, which are the same for both Alice and Bob, as a standard shared encryption key.

Armed with a deepening understanding of quantum communication, Bennett turned his attention to the problem of communicating a quantum state from one place to another without directly transmitting it nor even knowing what the state is. The catalyst for the solution to this "unknown quantum state transmission" came in 1991 from a paper by Asher Peres (1934–2005) of Technion University in Israel and Bill Wootters (1951–) at Williams College in Massachusetts.[40] Wootters was already famous for proving, with Wojciech Zurek (1951–) at Los Alamos, that quantum information cannot be cloned, known as the non-cloning theorem.[41] In the Peres and Wootters paper, they envisioned creating two identical but separated quantum states with the goal of ascertaining through successive measurement what the state is. They showed that partial information could be extracted if weak measurements could be taken and communicated back and forth to inform how to make the next measurements, and so on. However, they also showed that performing a joint measurement on both particles would always provide the most information. Wootters presented the paper at a conference with Bennett in the audience, and like the opportune meeting with Brassard 12 years before, when Bennett and Wootters got talking it launched something entirely new.[42]

Bennett was captivated by this process of making joint measurements on quantum particles (which is closely related to the HOM two-photon coincidence interferometry), and he sought to combine this procedure with his longstanding goal to communicate quantum states. In the ensuing days, a growing collaboration by email (at that time a new way of communicating among distant colleagues) expanded to encompass Brassard in Montreal, Peres at Technion, Claude Crepeau (1962–) in Paris, and Richard Josza (1953–) at Cambridge. Together, the group found the missing element that would allow an arbitrary and unknown quantum state to be "communicated" from one location to another without transmitting it directly and without ever knowing what the state was. This missing element was entanglement. They discovered that by introducing an entangled pair of states, and making joint measurements of the unknown state together with one of the entangled states, and subsequently communicating the results over a classical information channel, it was possible to convert the second half of the entangled pair into the unknown state. When it was time to name the new communication technique, Bennett suggested calling it teleportation (mixing Greek and Latin roots). Peres pushed instead for telephoresis (all Greek roots), but the team liked teleportation better, and that name stuck.

The protocol for quantum teleportation of a state $|\psi_1\rangle = \alpha |0\rangle + \beta |1\rangle$ with unknown coefficients α and β is as simple as it is profound. The essential resource that makes it possible is the nonlocal nature of entanglement of a two-particle state. Because of this nonlocality, operations performed by Alice on one half of an entangled pair are correlated to properties of Bob's other half of the entangled pair. The teleportation protocol is outlined in Table 9.1.

Table 9.1: Teleportation protocol.

Task:

- Alice seeks to transport a quantum superposition state $|\psi_1\rangle$ = $\alpha|0\rangle + \beta|1\rangle$ with unknown coefficients α and β to Bob without using a quantum channel.

Conditions:

- Alice cannot measure $|\psi_1\rangle$ directly nor clone it.
- Alice and Bob share an EPR pair $|\psi_{23}\rangle$ that they store until needed. (They get the pair prior to traveling apart.)
- Alice communicates to Bob using a normal (classical) channel.
- Alice should send as few classical bits as possible.

Approach:

- Alice entangles $|\psi_1\rangle$ with her half of the entangled pair by measuring a joint property of $|\psi_{12}\rangle$.
- Alice sends measurement results to Bob classically as bits.
- Bob manipulates his particle $|\psi_3\rangle$ depending on the measurement results he obtains, converting his particle to state $|\psi_1\rangle$.

The teleportation protocol, illustrated in Figure. 9.8, works because it induces a form of wavefunction collapse on $|\psi_3\rangle$ when the joint measurement on $|\psi_{12}\rangle$ is made, but it is only a partial collapse that gives no information about the coefficients α and β. In the simplest implementation, Alice measures one of four possible outcomes to her joint measurement (known as a Bell-state measurement) of the two particles with one outcome 0 or 1 for each particle. Alice then sends two bits of information to Bob as the two-bit strings 00, 10, 10, or 11. Bob has four possible quantum operations he can perform on his particle, and with the four bits of information he receives from Alice, he knows exactly which of the four operations will convert his state into the unknown state. This successfully teleports the state—without ever knowing what that state was.

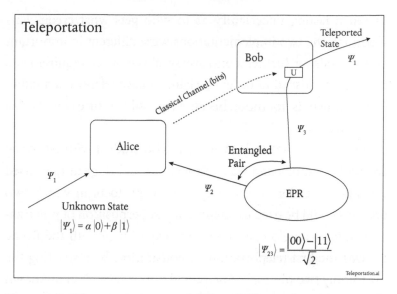

Figure 9.8 Teleportation protocol. Bob and Alice each get one particle of an entangled EPR pair $|\psi_{23}\rangle$. Alice has an unknown state $|\psi_1\rangle$ and measures a joint property of $|\psi_1\rangle$ and $|\psi_2\rangle$. The measurement result is sent to Bob as two classical bits who uses the information to manipulate $|\psi_3\rangle$ in one of four ways to convert it into the unknown state $|\psi_1\rangle$.

Two research groups raced to perform the first successful teleportation of an unknown state on a photonic qubit. One of these groups was spearheaded by Sandu Popescu (1956–) at Cambridge University collaborating with the experimental group of Francesco de Martini at the University Sapienza in Rome, Italy. The other group was led by Anton Zeilinger (1945–) of the University of Vienna, Austria. The Popescu group submitted their paper first in June 1997 followed three months later by the Zeilinger group in October. However, the Zeilinger paper was published first[43] in December 1997, while the earlier Popescu paper was delayed in review and was published with a 1998 publication date.[44] This flip-flop of submission and publication dates has caused a lasting uncertainty as to who gets what credit. Furthermore, the two implementations were different in important aspects, raising further controversy about what requirements need to be satisfied in order to claim to have teleported a qubit. The argument is not moot, because a possible future Nobel Prize could be at stake.

The approach taken by Popescu, who studied with Ahronov and with Shimony and was a Reader at Cambridge in 1997, used a modified version of the teleportation protocol in which two photons could be used to teleport any superposition of polarization states. Therefore, although the experiment with the Rome lab was the first teleportation demonstration, by restricting the use to only two photons, Alice could not be "given" an unknown state to teleport to Bob. The same entangled states that Alice and Bob shared were also used to establish the superposition to be teleported. Hence, Alice has only to look at the optical elements on her table to know what state is being teleported. This seeming nuance, in the minds of some, disqualified the Popescu experiment as literal teleportation. Although it represented a milestone in quantum optics and entanglement physics, establishing new techniques that advanced the field, it was not in the spirit of the Bennett protocol.

The Austrian group, led by Anton Zeilinger on the faculty of the University of Innsbruck, took a three-photon path to teleportation in the spirit of the Bennett protocol, and they were able to teleport a separate and unknown state. In the minds of many, this represented the first literal demonstration of teleportation. Ironically, the Zeilinger group was not initially looking for teleportation.[45] They had been exploring techniques for three-particle entanglement that could produce the state known as a

GHZ state[46] (named after Greenberger, Horne, and Zeilinger) and denoted as

$$|GHZ\rangle = \frac{|HHH\rangle + |VVV\rangle}{\sqrt{2}}$$

They hoped to use this state in higher-order tests of Bell's inequalities. (The H stands for "horizontal" and the V for "vertical" linear polarizations.) However, as they developed the supporting optical hardware and techniques, it became clear that they had nearly everything they needed to perform a strict teleportation demonstration. They took a quick detour, performed the experiment, wrote the paper, and submitted it to *Nature*, where it was published within two months, beating into print the Popescu publication that been submitted several months earlier. However, there was a fundamental limitation to their experimental teleportation protocol that persists to this day—it relies on optical Bell-state measurements that only give a usable answer some of the time.

This limitation on the optical Bell-state measurement is fundamental when using linear optical elements like mirrors and beam splitters and phase shifters. If there were ways to get the photons to interact in nonlinear optical elements, then all four Bell states could be distinguished. But nonlinear optics requires high photon fluxes, which is fundamentally incompatible with photonic qubits that come at a rate of only a pair at a time. Therefore, this Bell-state measurement has a fidelity of only 50 percent, leading to probabilistic teleportation, also known as *conditional* teleportation. Nonetheless, it is extremely simple, and when it works, it can be confirmed experimentally.

A simplified optical schematic of the Innsbruck experiment of Zeilinger and colleagues is shown in Figure. 9.9. The goal of this experiment was to confirm conditional teleportation of photonic qubits, so it dispensed with polarizing beam splitters in the Bell-state measurement, allowing it to only recognize one of the Bell states in the HOM interferometer. Another simplification was the use of post-measurement data analysis to verify teleportation rather than having Bob perform any manipulation of his half of the entangled pair. In this set-up, if the appropriate Bell state is detected, then the unknown state has automatically been transmitted, and all that is left is to confirm it by comparing the Bell-state detections with measurements on the teleported state.

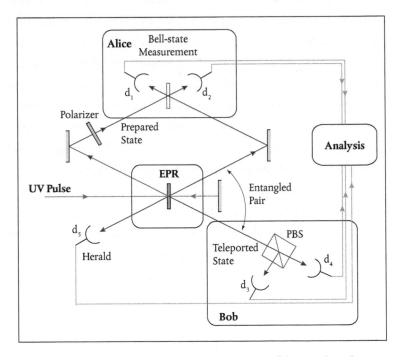

Figure 9.9 Quantum teleportation experiment of the Innsbruck group using a down-conversion crystal and a Bell-state measurement.

Today, quantum teleportation has moved far beyond these primitive demonstrations from 1997. Over the past two decades, teleportation of photonic qubits has spanned increasing distances, transmitted through fiber optics across the Danube in Vienna,[47] through 100 km of free space across a lake in China,[48] through 143 km across an ocean strait between the Canary Islands of La Palma and Tenerife,[49] and from Earth to a satellite.[50] During the same time, on the other end of the length scale, teleportation has been miniaturized to pass qubits between two circuit chips[51] as well as within a single chip.[52] The fidelity of teleportation has risen during the same time into the mid-90 percent range with efficiencies of several percent, while the formats available to optical teleportation have expanded to include continuous variables[53] (rather than discrete qubits), non-classical optical states called squeezed states,[54] and even "Schrödinger Cat" states.[55]

At the heart of quantum teleportation reside the dual principles of interference and entanglement. Entanglement provides the nonlocal resource that allows the manipulations and measurements on one particle to be correlated with manipulations and measurements on its distant twin. Interference provides a superposition of possibilities, and detection reduces the quantum superposition to specific outcomes, collapsing the output wavefunction into the wavefunction of the unknown state that was never looked at directly and hence remains unknown. Interference, in this quantum context, is the mixing of possibilities, a smorgasbord of possible paths for quantum particles to take. As Richard Feynman pointed out in his path-integral formulation of quantum physics, quantum particles take all paths available to them. And when they collapse by detection to a specific location, one does not say that the particle took one path or another—it is the action of taking all paths together that leads to the inter-ference. This interference, this mixing of possibilities, especially when the number of possibilities grows to exponential size, lies at the heart of the quantum advantage of information that is driving the next revolution in computing—quantum computing.

10

The Quantum Advantage

Interferometric Computing

> Quantum theory is a theory of parallel interfering universes.
>
> David Deutsch[1] (1985)

The secret behind almost every disruptive advance in the history of technology has been parallelism. Parallelism allows a system to do many things at once, cutting costs and time. Five hundred years ago, when Johannes Gutenberg conceived of movable type and the printing press, he parallelized the production of books, cutting the time to generate a full medieval manuscript from one in many months to many in one month. Two centuries later, the Industrial Revolution was built on the foundation of parallel production enabled by the newly harnessed power of steam. More recently, the speed of telecommunications rose exponentially after the introduction of fiber optics that packed an increasing number of parallel communication channels into a single fiber. Today, cloud computing and deep learning have emerged from parallel distributed processing, opening up new capabilities for computing power and artificial intelligence. These advances over the past half millennium have been driven by the

Interference. David D. Nolte, Oxford University Press. © David D. Nolte (2023).
DOI: 10.1093/oso/9780192869760.003.0010

parallelization of formerly serial processes. Now we are poised on the threshold of a new technological disruption based on parallelism that is unlike any that have come before—quantum computing.

Classical computers are intrinsically serial. They must execute each logical operation in sequence, completing one step before the next. Even parallelized computers, feeding multiple cores of central processing units with parallel threads of instructions, must wait for each thread to complete before assembling the partial results. In these classical systems, the time and cost of computing grows exponentially with the number of bits needed to provide an answer. In contrast, quantum superposition can provide a way out of this unfavorable scaling by replacing series of computations on individual bits with a single joint quantum measurement on all the bits at once. This is the promise of quantum computing—exponential speed-up.

Where does this exponentially large resource of information come from? It seems like quantum information pops into existence out of thin air, bypassing constraints on energy and entropy that are usually associated with information. In the Copenhagen interpretation of quantum physics there is no conflict: one would say that the information comes from quantum interference followed by wavefunction collapse. But there are others who would claim that the information flows into our universe and thus—through a kind of information conservation—it must flow out of others. This is the many-worlds interpretation of quantum physics. Although only a minority of physicists subscribe to this opinion, one of them harnessed this viewpoint to invent the first quantum algorithm.

David Deutsch's Interfering Worlds

John Archibald Wheeler (1911–2008) had an outsized influence on the history of twentieth-century physics. The child of librarians, he was a student of Niels Bohr, a colleague of Albert Einstein, a competitor of J. Robert Oppenheimer, and the mentor of Richard Feynman. During his tenure at Princeton, he wore neatly pressed suits and ties, spoke calmly and politely, yet pursued his physics like time was running out.[2] His main area of academic activity was in general relativity, and he literally wrote the book on it (with Charles Misner and Kip Thorne).[3] Yet his influence was often outside his main field of expertise, acting as provocateur, enabler, and nexus. He was a coiner of phrases, like "black hole" and "wormhole." He was a mentor and cheerleader for others whose ideas he championed or encouraged, even when he did not believe in the ideas himself but believed in the *people* making the theories.

One of Wheeler's graduate students at Princeton in the 1950s was Hugh Everett (1930–1982), a recent physics convert from chemical engineering whose plain outer appearance belied a vibrant internal mind and a hidden drinking problem.[4] Everett began by exploring the conditions needed to quantize general relativity (a task that has yet to yield a solution). As he worked, he realized that the foundations of quantum mechanics needed to be changed, because the Copenhagen interpretation of quantum mechanics, centered around Niels Bohr, seemed to pose an obstacle to progress. In an audacious step, Everett replaced the Copenhagen view with a self-consistent theory of his own that came to be known as the many-worlds interpretation (MWI) of quantum theory. This new view rid quantum mechanics of von Neumann's troubling concept of wavefunction collapse in one universe as Everett instead conceived of an infinite number of universes—a multiverse—among which all possible things occur. In the many-worlds view, the result of a quantum measurement is not random, but merely identifies which of the alternative universes the experiment resides in.

When it was time for Everett to defend his PhD thesis, Wheeler was torn between his respect for the self-consistent logic of Everett's many worlds and the conflict that it put him in with Bohr, his old mentor. To resolve the issue, Wheeler grabbed a final draft of Everett's dissertation and flew to Copenhagen in 1956 to consult with Bohr himself, hoping to convince him of the merits of the alternative interpretation. But the Copenhagen school had gatekeepers, Aage Peterson and Alexander Stern, who would brook no dissent from the Copenhagen interpretation. By the time Wheeler got Everett's thesis before Bohr, it was too late, and Bohr was unsupportive. Undaunted, Wheeler approved Everett's thesis and even arranged to have Everett spend time in Copenhagen, but Everett found it a hostile environment. Just as they had quashed Einstein's EPR objections, and Bohm's pilot wave, the monocracy in Copenhagen rejected his many worlds. Discouraged, Everett left physics and spent the rest of his career at the Pentagon.

Despite the lack of support from the Copenhagen school, the MWI refused to die. It had a tendency to rear its head every time the bizarreness of quantum physics became too much to bear for some faction of physicists. For instance, the MWI went through a renaissance in the 1970s as physicists began grappling with the phenomenon of decoherence—the mechanism by which quantum systems lose their pure quantum superpositions and become statistical mixtures of measurement outcomes. Around that same time, Wheeler received a phone call from an unknown British graduate student who was visiting the States and wondered if they could meet to talk. Wheeler loved talking, and would talk to anyone about anything, but his time was precious, so he tended to set up meetings on the fly—literally. Without thinking whether this British student needed to be in Maine or not, he suggested meeting at the airport in Portland and flying together to Boston.

David Deutsch (1953–), the British student from the UK who was studying in Oxford, dutifully went to Maine in 1977 to meet Wheeler. They talked nonstop for an hour in the airport, on their feet the whole time, and continued talking on the plane.[5] When they parted in Boston, Deutsch thought that was the last he would see of Wheeler, but a year later, while he was standing in the dinner line at a conference, Wheeler walked up to him and suggested that he should come to visit him at the University of Texas at Austin, where he had recently moved from Princeton. Deutsch agreed and was soon immersed in the same topic that had caused Everett to formulate his many worlds— the problem of quantum fields in curved space-time—and just like Everett, Deutsch ran into obstacles associated with von Neumann's wavefunction collapse. Deutsch became a convert, convinced that Everett's multiverse was the true source of quantum weirdness—he only needed to prove it. Enter Wheeler as nexus, a gatherer of minds, with a keen insight into what's next.

At that time, Wheeler had become interested in the physics of information, especially information that falls into a black hole, and was developing ideas about how physics and computing shared fundamental origins, so he arranged a small conference and invited a select group of physicists to discuss the physics of computing. Deutsch attended, only half listening. That evening, while talking with Charles Bennett, Deutsch was less than awed with what he had seen at the conference and said so to Bennett, calling some of it "rubbish." Bennett was polite and asked him why he thought so. Deutsch replied that, "there's nothing fundamental in the theory of computation that tells you what the instruction set is." Bennett paused, then said simply, "The instruction set is physics."[6]

It is sometimes hard to fathom what triggers a new idea, but Bennett's simple response to an impolite prod from a somewhat peeved Deutsch created a sudden crystallization in Deutsch's mind. Deutsch recalled that he actually gasped and took two steps backward. With a clear vision over the next few years, Deutsch worked to retool the theory of computation based on the principles of quantum theory. In particular, he had glimpsed how to use the idea of a quantum computer to prove the existence of the multiverse. The key was interference—not the interference of the Copenhagen school in which a photon in a double-slit experiment takes both paths—but where the photon goes one way in one universe and another way in another universe, and the two universes themselves interfere. An interferometer is where two worlds make contact, and Deutsch believed that, if he could construct a quantum algorithm that operated on more information than could exist in each universe separately, then he would have his smoking-gun proof.

The Proof Is in the Algorithm

A central operation of a classical information system is the evaluation of a function. In a program or inside a computer chip, a function evaluation is a computational step, mapping an input to an output. Even the simplest of all computations, namely $b = a$, is a mapping of a value "a" onto "b." If a function operates on bits of information, then one bit can have two possible values, and the function must be evaluated twice to sample all possibilities. If the function operates on two bits, then the function must be evaluated four times to exhaust all possibilities. This is simple exponential scaling in the number of bits. For instance, a function on N bits has 2^N values. The bottleneck in classical information flow is the isolated function call that operates bit by bit.

After Deutsch met Bennett at the cocktail party at Wheeler's house in Austin, and as he began constructing a theory of computation based on quantum mechanics, Deutsch entertained the idea that information from many worlds could be tapped in a single step—all the different function values accessed at the same time, each from a different universe. But even in the many worlds interpretation of quantum physics, one cannot see all values separately—only one measurement outcome at a time is allowed, even if all the information is there, because the measurement is simply telling us which one of the many possible universes we reside in. Therefore, Deutsch searched for a new type of algorithm that used a joint property of the function across all the universes to arrive at a single answer, and he found it.

His first quantum algorithm was merely a toy—simple and useless for any real-world application. But it accomplished what Deutsch needed—finding a joint property of a function in a single evaluation. The question the quantum algorithm answered was whether two queries of a two-valued function have the same values or different values; that is, whether it is a "constant" function or a "balanced" function, respectively. Denoting the two queries as

0 and 1, then the two function values can satisfy f(1) == f(0) in the constant case, or f(1) ~= f(0) in the balanced case. These represent two possibilities, but since the functions are binary, there are four classical states: in the constant case, either f(0) = 0 and f(1) = 0, or f(0) = 1 and f(1) = 1; in the balanced case, either f(0) = 0 and f(1) = 1, or f(0) = 1 and f(1) = 0. In all, there are four possibilities that require four classical evaluations. The joint property that Deutsch chose for his algorithm is the bitwise AND (expressed by the symbol $\oplus$). Then for both constant cases $f(0) \oplus f(1) = 0$, while for both balanced cases $f(0) \oplus f(1) = 1$. Note that four possibilities are reduced down to two. If the function is constant, then the joint property equals zero. If the function is balanced, then the joint property equals one. Furthermore, because quantum mechanics allows for superpositions, a single wavefunction can contain both function evaluations. Therefore, Deutsch had found a quantum algorithm that performed twice as fast as a classical one.

A schematic of the Deutsch algorithm, including the wave-functions at different locations through the quantum gates, is shown in Figure. 10.1. The wavefunction is shown at each stage of the computation moving from left to right. The two initial Hadamard gates, denoted by the H in the small square box, create superpositions in state $|\Psi_1\rangle$ shown as the mixing wavefunction in the figure. Then the function evaluation in the large square box produces two possible superpositions from the input, and the final (single) measurement after the final Hadamard gate gives $M = f(0) \oplus f(1)$. If M = 0, then the function is constant. If M = 1, then the function is balanced.

When thinking of implementing this algorithm, it is important to keep in mind what resources need to be expended to implement it. Any quantum algorithm can be implemented on a

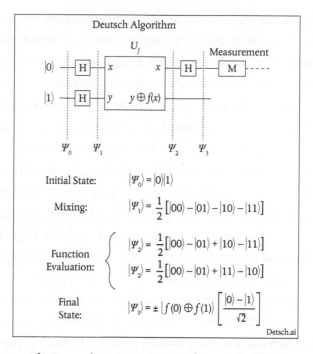

Figure 10.1 The Deutsch quantum circuit diagram for finding the constant/balanced property of the function. The wavefunction is shown at each stage of the computation moving from left to right. The function evaluation produces two possible superpositions, and the final (single) measurement gives $M = f(0) \oplus f(1)$. If $M = 0$, then the function is constant. If $M = 1$, then the function is balanced.

classical computer, but the resources it takes, whether of time or of parallel channels, grows exponentially with the number of bits. For a quantum computer, on the other hand, the resources needed to perform the algorithm would grow roughly as a polynomial in the number of bits. This was the key. Deutsch's algorithm could in principle be executed on a quantum computer with exponential speed up—this was the smoking gun he needed to support the many-worlds interpretation.

By 1985 Deutsch was back in Oxford when his paper "Quantum-theory, the Church–Turing principle and the universal quantum computer" was published in the *Proceedings of the Royal Society*.[7] As usually happens when an earth-shaking paper is published, almost nobody noticed. The purpose of the paper was closer to quantum philosophy than to computer science. In Deutsch's opinion, the existence of this first quantum algorithm was a proof of the existence of multiple universes, because the advantage of quantum computation was coming from the interference of information among multiple worlds. However, one person's proof is another person's counter-proof, and the Copenhagen interpretation of quantum mechanics has no trouble explaining the advantage. For Copenhagen, the joint property is an interference among the different quantum amplitudes of a simple quantum superposition. Therefore, in the battle between the Copenhagen and the many-worlds interpretations, the paper was a draw.

Deutsch's search for quantum algorithms was happening in the background of a more mainstream interest in quantum computing that had grown from a lecture given by Richard Feynman at a conference held in 1981 on the physics of computation, hosted outside Boston by IBM and MIT. Feynman's presentation on "Simulating Physics with Computers" was published a year later, in which he outlined the essential need for quantum hardware to efficiently simulate the properties of quantum systems. His ideas were fully within the Copenhagen tradition and hence were more palatable to most physicists, and his suggested applications for quantum computers were inherently practical. Calculating many-body properties of quantum systems was prohibitive using conventional computers, but he showed that quantum simulation on quantum computers could be done exponentially faster. Possible outcomes of such quantum calculations might be improved chemicals or pharmaceuticals, making his arguments in favor of quantum computing compelling.

Therefore, for the practical-minded, whether the existence of a quantum algorithm proved or disproved the existence of multiple worlds was largely moot, partly because the quantum algorithm does not perform an exponential number of classical computations, nor does it give an exponential number of answers. It gives a single answer in which an exponential number of contributions interfere. Although Deutsch may have failed to win many scientists over to the many-worlds view, his paper did introduce the world to the first of what would become a widening range of quantum algorithms—and some of them would turn out to be just as important as chemical calculations or more so. The key to these algorithms is intrinsic quantum parallelism that relies on the interference of many channels of coherent information.

As a first step down this road of quantum algorithm development through interference, Deutsch published an extension of his algorithm in 1992, with the Australian mathematician Richard Jozsa (1954–),[8] taking the original single-bit function and extending it to N-bits. In the Deutsch–Jozsa version, the function either has all the same values (constant), or it has equal numbers of ones and zeros (balanced). The joint property is again either an output of zero if the function is balanced or an output of one if it is unbalanced. The exponential speed-up of the quantum algorithm is clearer in this N-bit situation. In the classical case, to tell whether the function is constant or balanced would take at worst $2^N/2+1$ function evaluations, because the first half of the queries could all be the same for either function case. Then the very next query would tell if the pattern is broken. For large N that is approximately $2^{(N-1)}$ function queries. But the quantum algorithm would require only a single evaluation—clearly an exponential speed-up.

There is a simple optical analog that can implement the Deutsch–Jozsa algorithm by using a single function evaluation on N channels to decide if the function is balanced or not. The analog relies on the properties of diffraction and interference, and the optical hardware uses just a single lens and an optical phase modulator, shown in Figure. 10.2. The N-bit function is implemented by the optical phase modulator that induces a zero or a pi phase shift on transmission through each pixel, serving as the bits that are zero or one. A laser beam illuminates the "qubit" plate

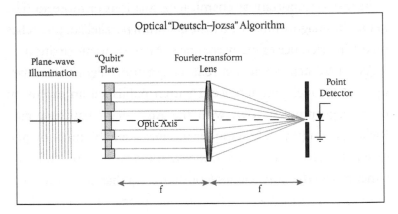

Figure 10.2 The Deutsch–Jozsa algorithm can be demonstrated classically using a simple phase modulator, a lens, and a photodetector. If the modulator is uniform, then constructive interference is measured at the detector. If the modulator has equal numbers of pi and zero phase shifts, then by the principles of Fourier optics, net destructive interference occurs on the optic axis and no light is measured at the detector.

that diffracts the beam into multiple waves at multiple angles as a diffraction pattern. A lens captures the scattered waves and directs them to a detection plane where an aperture has a small hole on the optic axis that selects the zero-order diffraction of the phase plate. If the function is constant, all the partial waves at the detector add in phase to constructive interference, and a high signal is detected. If the function is balanced, with equal numbers of ones and zeros, then all the contributions add out of phase (equal number of zero or pi phase shifts) to destructive interference, and no signal is detected. A single function evaluation (the illuminating laser beam and the detected signal) serves, through *classical* interference, to answer the question of balance.

This simple optical (and classical) implementation of the Deutsch–Jozsa algorithm serves three didactic purposes. First, it shows how the key element of interference allows a joint property to be measured in a single operation. Although each partial wave, or each component of a superposition, carries a little information, the algorithm does not need to know all the individual values—it just needs to know how they compare on the whole, adding up through constructive or destructive interference. This is a common feature in a wide range of quantum algorithms today, even in (or especially in) the more sophisticated algorithms that do have real-world applications (more about this in a moment).

The second didactic purpose of this simple optical Deutsch–Jozsa algorithm is to demonstrate how classical hardware can implement a quantum algorithm, because all quantum algorithms can be simulated on classical hardware. In fact, classical implementations of interference effects are part of *classical optical computers* which can provide substantial speed-up of performance relative to their electronic counterparts. For instance, there are hardware companies today that are developing special-purpose optical-based accelerators used in server farms where they provide much faster and more energy-efficient computation in cases where optics has a clear advantage over digital circuitry.[9]

Finally, the third didactic purpose of this example is the illustration that exponential resources are needed to perform the classical implementation of the quantum algorithm. The spatial phase modulator is driven by a conventional computer interface that requires a separate function evaluation to address each pixel of the modulator, so no parallel advantage is gained in this classical computation. For the quantum algorithm to have computational advantage, it needs to be implemented on quantum hardware. But first, it is necessary to have an algorithm worth all the effort of developing quantum computers—a killer app. This was provided by a quirky, wild-haired, and at the time unknown, computer scientist at Bell Labs, whose quantum app threatened to hack every secure encoded transaction on the planet.

Peter Shor's Periodic Program

The reason why you don't hesitate to enter your credit card information to buy products online comes down to a surprisingly simple asymmetry in the multiplication and division of numbers. If I give you two prime numbers n_1 and n_2, you can multiply them together as $n_1 \times n_2 = N$ using a common computer in mere microseconds. But if I give you the product N of two prime numbers, and if the prime numbers n_1 and n_2 are large, then it might take your computer an hour to find the prime factors. And if those numbers are very large, let's say each one is 256 bits long, then it will take your computer years or longer to find the answer. This is known as the asymmetry of prime factoring. It takes almost no time to multiple two large numbers together, but it takes a very long time to factor a large product into its individual prime factors. In fact, the time and resources needed for prime factoring grow exponentially in the number of digits of the prime factors, making the problem intractable if the numbers grow too big.

This exponential size of the prime factor problem is the basis for the most widely used encryption protocol on the internet. The encryption protocol is called RSA, named after Ron Rivest, Adi Shamir, and Leonard Adleman, who wrote about the scheme in 1977. It didn't matter that they published their work openly,[10] because the strength of the RSA encryption lies in the hard problem of factoring. Knowing that an encrypted message is using RSA provides almost no help to an eavesdropper. The receiving party sends their public key, consisting of the product of primes plus an additional integer that depends on the prime factors, to the sender. The sender can use these two numbers to encrypt a message to send it back—even broadcasting it widely if they wish. The sender never knows what the prime factors are, but the encrypted message can only be decoded by using the prime factors, and only the originating party knows what they are. This is the ingenious part of public key distribution (PKD). The key is sent openly, and even the encrypted message can be sent openly, but almost no one can decode it, unless they have a very powerful computer that can factor the product, or better, a computer with exponentially more resources than ordinary computers.

In 1993, Peter Shor (1959–) was an early career computer scientist working at Bell Labs in New Jersey when he was asked to serve on the review committee for an important international conference on theoretical computer science known as STOC (Symposium on the Theory of Computing). He had heard of quantum computing through weekly seminars held at Bell Labs, but so far he had not been impressed with what they could do, because only toy problems (like Deutsch's algorithm) had been proposed and solved using quantum algorithms. Yet one of the abstracts that he reviewed was from Daniel Simon, a PhD student of Gilles Brassard, who had developed a quantum algorithm that could find periodic patterns in high-dimensional lattices. Shor was immediately struck by the general power of the approach, which used something called a quantum Fourier transform that had recently been developed by Umesh Vazirani at UC Berkeley. Shor argued for the acceptance of Simon's abstract to the conference, but the symposium committee voted him down, and the abstract was rejected. Nonetheless, Shor began thinking about extensions of Simon's algorithm to a different problem that had hidden periodicities, known as discrete logarithms.

Not wishing to be a parasite (someone who secretly reviews someone else's work and appropriates it as their own), Shor contacted Simon directly and asked for a preprint of his work to have an official version to work from. Within a few months, Shor had developed a quantum algorithm to solve the problem of discrete logarithms using the quantum Fourier transform, which was an exponentially hard problem in general, even though there were special cases that could be found quickly. He prepared to show his findings in April 1994 at the weekly Bell Labs seminar, which was famous for giving speakers a hard time. (When I was a post-doc at Bell Labs in the late 1980s, the weekly seminar notice included a cartoon picture of a pack of hungry wolves sitting in chairs surrounding a frightened and sweating speaker who looked like lunch.) Shor's talk went well, and he mentioned in passing that discrete logarithms could be used for encryption. He knew that prime factoring was related to the problem of discrete logarithms, but he hadn't started on that problem, and he wanted to keep that close to his chest so that he wouldn't be scooped by someone else.

Yet somehow, as word spread, inside Bell Labs and out, that Shor had cracked the discrete logarithm problem, it morphed into the story that he had cracked RSA. Therefore, Shor was shocked when, five days later, he received a phone call from Vazirani who told him, "I hear you can factor on a quantum computer, tell me how it works."[11] Fortunately, in those same five days, Shor had actually cracked RSA using the same quantum Fourier transform, so he could tell him. The key to the algorithm came from number theory in which a property of the product of primes produces long sequences of numbers that have built-in periodicities. If those periodicities could be identified (a problem that was exponentially large in the number of digits of the prime factors), then by running the algorithm many times, there was a finite probability that the prime factors could be found in a time that scaled as a polynomial (not exponentially) in the length of the prime factors. The algorithm cleverly used the asymmetry of the multiplication and division of integers, because any given cycle of the algorithm was most likely to give the wrong factors, but these could be checked in microseconds on a classical computer, and if wrong, then the quantum algorithm could be run again and again, many times, until the right factors were produced.

The Shor algorithm for prime factoring uses quantum superposition and interference to identify the periodicities associated with products of prime numbers. The reason the periodicities exist in the first place is an arcane aspect of number theory whose explanation will be left to number theorists. The point is that periodicities in functions are accessible to Fourier transforms. Once again, a classical optical analog suggests itself as an intuitive way of thinking about the Shor algorithm—the optical diffraction grating in Figure. 10.3. When a grating with a periodicity Λ is illuminated by a plane wave, the grating generates multiple

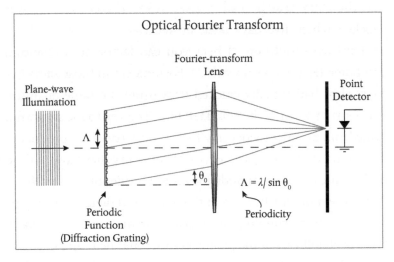

Figure 10.3 Classical optical analog of a quantum Fourier transform. The function values are represented by the diffraction grating. For a specific angle θ_0 related to the periodicity Λ of the function, all the scattered partial waves interfere constructively and are detected in a single point measurement. The location of the detector is chosen randomly, as is the argument x of the modulus function $f_N(a) = \text{mod}(x^a, N)$ that generates the diffraction grating. Most of the time the detector fails to measure a peak, but when it does, then the period Λ is found, and the product N is factored. This classical analog is not itself a quantum computer, because the diffraction grating must be addressed using all the conventional function calls. A quantum computer would represent the diffraction grating with a superposition of qubits.

scattered partial waves that propagate away from the grating at all angles—this is just Huygens' principle. But because of the periodicity of the grating, there are only certain angles where the phases of all the partial waves add up in constructive interference to produce a diffraction spot. At all other angles, the partial waves cancel each other out through destructive interference. To find the periodicity of the grating, all one has to do is find the angle θ of the diffraction peak and do the simple math to find $\Lambda = \lambda / \sin \theta$, where λ is the wavelength of the plane wave.

The Shor algorithm does essentially the same thing. A modulus function is defined based on randomly picked integers x and a that have certain simple relationships to the product N that is to be factored where $f_N (a) = \mod (x^a, N)$. This function has an unknown period Λ that depends on x and a. By encoding this function onto qubits, a quantum Fourier transform creates a superposition of all the partial function values (just like all the different positions on a diffraction grating) and then a single measurement is made (just like measuring the intensity of the diffraction pattern). Most of the time the measurement yields nothing because of destructive interference, like picking the wrong angle to measure the diffraction pattern. But eventually, by random chance, the right values of x and a will be picked that produce constructive interference. Once this happens, it is simple math to find the periodicity Λ. And once the periodicity is found, then the product $N = n_1 \cdot n_2$ can be factored using conventional computers. The answer is

$$n_1 = GCD \left(x^{\Lambda/2} - 1, N \right)$$
$$n_2 = GCD \left(x^{\Lambda/2} + 1, N \right)$$

where GCD stands for greatest common divisor.

The surprising part of the Shor algorithm is that it usually fails, because the wrong x or a were picked, and there are far more wrong values than right values. Yet the time spent finding wrong answers before the right answer is found is exponentially shorter than if one were just randomly guessing prime factors directly. The exponential speed-up comes from superposing all the function values together and then making a single measurement of the resulting interference.

After Shor officially announced his quantum algorithm for prime factoring,[12] the flood gates opened. The RSA encryption scheme is used for a large fraction of the secure communications in the world, ranging far beyond credit card purchases to bank transactions and even secret government communications. Whoever could get their hands on a quantum computer to run Shor's algorithm could read the secret communications of the world like an open book. Of course, the US National Security Agency (NSA), the largest organization of code breakers in the country, took an immediate interest, as did funding agencies. Quantum computing was suddenly relevant—not just some esoteric toy for physicists and computer scientists to play with.

Quantum Interferometric Logic

Once the idea of a useful quantum algorithm had been launched into the broader world, interest in quantum computing increased dramatically in the mid-1990s. Despite the fact that the foundation of the new field rested on arcane aspects of quantum physics, the growing subject drew from the mature fields of computer science and electrical engineering which were busy at that time making increasingly smaller logic gates and packing them ever more tightly onto integrated circuit chips. By analogy, if quantum bits of information were to be processed in a circuit, then there must be the quantum equivalent of transistors and other logic "gates" that also could scale to the level needed to perform useful quantum algorithms. This gave rise to the quantum circuit paradigm of quantum computing, where quantum bits of information travel through successive gates, and qubits can control the states of other qubits. It was quickly discovered by David DiVincenzo, Adriano Barenco, Harald Weinfurter, and Seth Lloyd around 1995 that only a few types of gates needed to be daisy-chained together to create a universal computer; that is, a computer that can compute any finite algorithm.

The idea of using optics and photons to build quantum logic was an early favorite because photons were already being used for quantum teleportation. Teleportation could serve as the communication bus of a quantum computer, and the coherent properties of light are easily controlled by the use of lasers and optical modulators. Mandel's work, in particular on single-photon optics and beam splitters, was already performing a type of logic if the input and output paths (or modes) of a beam splitter were viewed as separate "rails" encoding logical qubits. The dual-rail representation of a beam splitter relies on single-photon manipulation, where a single photon is in one, or the other, of the two input paths into the beam splitter. The two input paths play the role of the two states of a qubit. If the photon is in the "first" path, then the qubit state is $|0>$, and if the photon is in the "second" path, then the qubit state is $|1>$. The same representation applies to the two output paths. The beam splitter performs a unitary operation (an operation that does not modify the total probability but only acts as a general

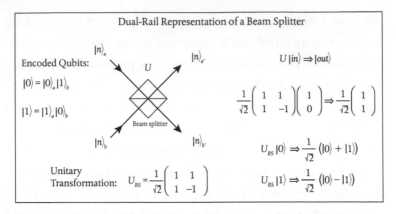

Figure 10.4 Dual-rail representation of a quantum beam splitter operating on a single photon as a unitary matrix. The qubit encoding for $|0\rangle$ is zero photons in the upper mode and one photon in the lower mode. The qubit encoding for $|1\rangle$ is the flip with one photon in the upper mode and zero photons in the lower mode. The beam splitter has an equivalent matrix operator U_{BS}. The mode occupancy n can only be zero or one in this dual-rail representation.

rotation matrix operating on the qubits) on the input modes, creating superpositions in the output modes, shown in Figure. 10.4. The unitary matrix of the beam splitter is also equivalent to a logic gate known as a Hadamard gate.[13]

The beam splitter is a single-qubit gate, and single-qubit gates can be daisy-chained to form additional gates. For instance, two beam splitters in sequence form a Mach–Zehnder that has the properties of a logical NOT. However, the NOT gate is also a one-qubit gate, and universal quantum computing requires at least on logic gate in which a qubit controls another qubit. One such two-qubit control gate that is central to the quantum gate paradigm is the controlled-NOT or CNOT gate, shown in Figure. 10.5. The qubit on the control line passes through unchanged,

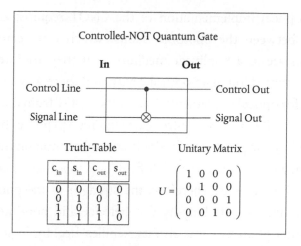

Controlled-NOT Quantum Gate

Figure 10.5 The Controlled-NOT gate, known as C-NOT, is a reversible gate in which the NOT function is applied to the signal line if the control line is "1," and does nothing otherwise. The truth table and unitary matrix for the gate are also shown.

while the qubit on the signal line is affected by the value of the control qubit: if the control qubit is a logical |0> then the signal qubit passes through unaffected; if the control qubit is a logical |1>, then the signal qubit is "flipped" such that |0> becomes |1> or |1> becomes |0>. These actions are shown in the Truth Table in the figure. The gate can also be described by a 4 × 4 unitary matrix. The power of the CNOT comes when the input qubits are quantum superpositions. For instance, if the control is a balanced superposition of |0> and |1> and the signal input is simply |0>, then the action of the CNOT is

$$\frac{1}{\sqrt{2}}\left(|00\rangle + |10\rangle\right) \rightarrow \frac{1}{\sqrt{2}}\left(|00\rangle + |11\rangle\right)$$

creating an entangled state that can be used for teleportation or as a resource for quantum algorithms.

An optical implementation of the CNOT requires an inter-action between the photonic qubits, which can occur when photons are in a nonlinear medium, or if they are interacting with a trapped ion in an optical cavity, and there was an intense search for optical implementations of the CNOT from the around mid-1990s. But an ideal control gate requires a pi phase shift to be induced by the presence of a single control photon on the signal photon, which is a severe and difficult goal to achieve using two-photon interactions. Therefore, the strength of the photon for communication—its long-lived coherence and non-interaction with other photons—was also its Achilles' heel for logic. By the turn of the millennium in 2000, it looked like all-optical systems would not be a viable technology for quantum computing. But then, to the surprise of many, the spooky properties of quantum measurement came to the rescue, opening the door to a relatively easy implementation of quantum logic using photons—as long as one was willing to turn away from the entrenched mindset of deterministic quantum computing.

Linear Optical Quantum Computing

The most important take-away message from quantum infor-mation theory is that quantum measurements are nonlocal. This means that when multiple information streams become entan-gled, and when the information on one stream is measured, it "collapses" the information on the other streams to states that are correlated with the measured values. This is why the CNOT has unintuitive (when viewed classically) properties. In the circuit diagram in Figure. 10.5, the control line goes through the gate unaltered, and so by classical logic, there should be nothing you can do to the signal output that should affect the control output. But quantum mechanically, the gate entangles the control and signal, so that measuring the signal output collapses the control output wavefunction. This is the nonlocal property of the CNOT gate, and nonlocal properties of quantum gates (even linear gates that do not need photons to interact directly) can be used to make a new type of quantum logic.

On July 24, 2000, a landmark paper was received by *Nature* magazine submitted by Emanuel Knill and Raymond Laflamme at Los Alamos National Lab in the United States with Gerald Milburn from the University of Queensland, Australia. This paper proposed a technique, now known through the acronym KLM, that could perform quantum logic in a universal way using only linear optical elements like single-photon sources, beam splitters, phase shifters, and single-photon detectors.[14] This paper was revolutionary for two reasons. First, it released the fetters that optical nonlinearity had clamped onto optical quantum computing. Second, it enlisted the intrinsic nonlocality of quantum measurement to perform quantum logic, introducing the novel idea of "measurement-based" quantum computing.[15]

The central idea of linear optical quantum computing is that rather than "control" qubits, one introduces many extra qubits, known as ancilla qubits, that are entangled with signal qubits. In a single "instance" or "try" of the computation, all the photons are sent through the optical system and are detected by photodetectors at the outputs—both signal and ancilla qubits. However, if (and only if) the ancilla qubit measurements come out a specified way would a useful computation have taken place on the signal qubits. If some other outcome is measured on the ancilla detectors, then that try is thrown out and another set of photons is sent through the system. This can be done many times until by chance the ancilla measurements come out just right, and then the signal photons will be known to have done the correct quantum computation even though they themselves were never measured. The KLM authors called this process *post-selection*—one tries the computation many times, but post-selects only for the instances when the ancilla measurements guarantee the right outcome for the signal qubits. The ingenious part of this new type of quantum computation was recognizing that the photodetection process on the ancilla photons provided an effective nonlinearity through wavefunction collapse that took the place of an actual nonlinearity. The authors then showed how a universal set of linear operations, with post-selection, could be universal for quantum computation.

At the same time that KLM schemes were revolutionizing the design principles of optical quantum computers, integrated photonics was maturing as an integral part of fiber-optic telecommunication systems. Photonic integrated circuits (PICs) are the optical analog of electronic integrated circuits where photons replace electrons, waveguides replace wires, and Mach–Zehnder interferometers replace transistors. The advantage of putting photonics onto chips, instead of using free-space optics or fiber optics, is the vast silicon miniaturization and integration technology that has been developed over the past half century for electronic computer chips. Just as more and more transistors are being placed on silicon chips, more and more interferometers can be integrated into compact packages on PICs. Furthermore, the requirement of high-quality fabrication and low losses for telecom applications of PICs are the same requirements for photonic quantum chips. Therefore, photonic quantum chips have become the workhorses of KLM and related schemes of linear optical quantum computing.

An example of an integrated Mach–Zehnder (MZ) interferometer is shown in Figure. 10.6. The lines represent ridge waveguides fabricated as silica or nitride oxide strips on a silicon chip. When two waveguides come very close together, they couple the electric fields of the light between the two waveguides and act as a beam splitter. Two such beam splitters in sequence makes an MZ interferometer. The interferometers are controlled (programmed) by phase shifters that are fabricated over the waveguides. These can be as simple as little local heaters that are driven by conventional currents from a computer station. The external computer controls the heating and hence the phase shifts of these phase shifters, allowing external programming control of the operation of the interferometer. If the input is a single photon, then the output of the interferometer is a superposition state, and the interferometer acts as a Hadamard unitary gate on the dual-rail input.

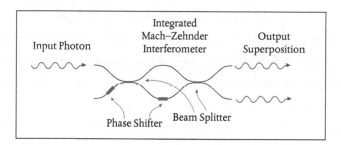

Figure 10.6 An integrated Mach–Zehnder interferometer constructed of ridge waveguides and phase shifters on silicon. When two waveguides come very close, they split the light between the two waveguides, acting like a beam splitter. When the input is a single photon, the output is a quantum superposition.

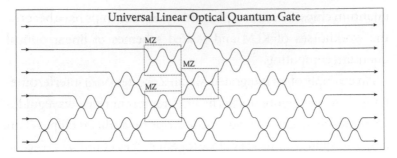

Figure 10.7 An example of a 64-qubit universal quantum gate composed of a cascade of 15 Mach–Zehnder interferometers (several individual MZs are shown in the dashed boxes).

An MZ acts as a transistor, and just as in integrated circuits with many transistors, many of these interferometers can be connected together on a photonic quantum chip. An example [16] with 15 MZs is shown in Figure. 10.7. The multi-mode interferometer in the figure has six input and output waveguides that can represent $2^6 = 64$ states. The overall operation of the 15 MZ interferometers (each with an addressable phase shifter [not shown]) is a general 6×6 unitary matrix operator. This circuit can be programmed to perform any generalized rotation on the six qubit states. Chips such as this are just the beginning. Moore's law of silicon chips, introduced by Gordon Moore (1929–) of Intel corporation in 1965 as the doubling the number of integrated components on chips every two years, had a good run for nearly half a century, and PICs are just beginning, so a new Moore's law for integrated photonics is already underway.

Nonetheless, despite the positive prospects of linear optical quantum computing and the polynomial scaling of resources that it promises, the number of required resources is still very large and remains an obstacle to the implementation of universal quantum computers using light and interference. A partial solution to this problem may come if one relaxes the requirement for a *universal* quantum computer. After all, there are many special-purpose electronic chips inside computers that do a wide variety of important functions, such as graphical processor units (GPUs). Similarly, there are important subsets of quantum applications that can be solved exponentially quickly by arrays of non-universal linear interferometers.

Boson Sampling and the Quantum Advantage

In computer science, a *complexity class* categorizes algorithms by how many steps they take to solve a problem, and a "zoo" of computational complexity classes[17] has been assembled over the years, containing hundreds of different classes with an even larger number of relationships among them. Exploring these relationships helps keep an army of computational theorists gainfully employed, because questions about complexity class are not esoteric—they pertain to how easily algorithms can be implemented in practice, or whether they can be implemented at all. After the publication in 2001 of the KLM post-selection protocol using quantum optical linear gates, there was an immediate search for the kinds of problems it could solve, and questions were raised about which computational complexity class such non-deterministic algorithms belonged to.

The theoretical computational scientist Scott Aaronson (1981–) worked on the problem of the computational complexity class of the KLM protocol for his PhD in 2004 in quantum computational theory with his advisor Umesh Vazirani at UC Berkeley. His conclusion was that the problems it could solve coincided with the classical complexity class of probabilistic polynomial-in-time algorithms (known as PP). This was an important result, bolstering the importance of linear optical approaches, because it showed again (in a different way than the Shor algorithm) that quantum algorithms escaped the exponential resource bottleneck of classical computers. After graduating from Berkeley, Aaronson took a post-doc position at the Institute of Advanced Study in Princeton and then a position at MIT. During that time, he continued to explore what other complexity classes could be associated with linear optics. Linear optics has always been the most robust way of maintaining quantum coherence in laboratory systems, and hence it was a prime contender for an eventual quantum computer. In 2011, Aaronson, with his post-doc Anton Arkhipov at MIT, published a landmark paper titled "The Computational Complexity of Linear Optics." They speculated on whether there could be an application of linear optics *without* post-selection that could still demonstrate quantum computational advantage, solving problems intractable on a classical computer. The answer was something they called "boson sampling."

To get an idea of what boson sampling is, and why it is hard to do on a classical computer, think of the classic demonstration of the normal probability distribution found in almost every science museum you visit, illustrated in Figure. 10.8. A large number of ping-pong balls are dropped one at a time through a forest of regularly spaced posts, bouncing randomly this way and that until they are collected into bins at the bottom. Bins near the center collect many balls, while bins farther to the side have fewer. If there are many balls, then the stacked heights of the balls in the bins map out a Gaussian probability distribution. The path of a single ping-pong ball represents a series of "decisions" as it hits each

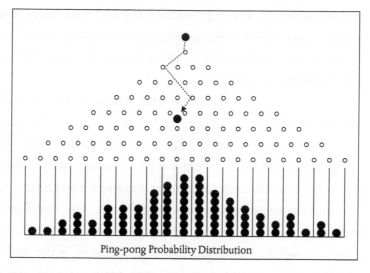

Ping-pong Probability Distribution

Figure 10.8 Ping-pong-ball demonstration of the normal probability distribution. Each dropped ball makes a decision to go left or right at each post it hits.

post and goes left or right, and the number of permutations of all the possible decisions among all the other ping-pong balls grows exponentially—a hard problem to tackle on a classical computer.

The core computational feature of the ping-pong problem is something known as the *permanent* of a matrix (in contrast to the *determinant* of a matrix). The permanent of a matrix is the recursive product of its elements with the permanents of its submatrices—just like a determinant but *without* alternating signs. Determinants larger than 3 × 3 are already something we don't like doing by hand. Beyond that, we resort to a computer. If the matrix is N-by-N, where N is a very large number (let's say in the hundreds), then even a computer will bog down if it doesn't resort to well-known algorithms for determinants. The algorithms rely on the fact that signs alternate when taking a determinant. But for a permanent of a matrix, without the sign flips, there is no efficient algorithm—it belongs to a computational complexity class known as #P which is one of the more difficult complexity classes (harder even than NP).

In their 2011 paper, Aaronson and Arkhipov considered a quantum analog to the ping-pong problem in which the ping-pong balls are replaced by photons, and the posts are replaced by beam splitters. The simplest possible implementation could have two photon channels incident on a single beam splitter—the HOM interferometer of the previous chapter. The well-known result in this case is the two-photon "HOM dip," which is a consequence of the boson statistics of the photon. In the Aaronson scheme, this system is scaled up to many channels for an N-channel multi-photon HOM cascade, shown in Figure. 10.9. The output of this photonic "circuit" is a sampling of the vast number of permutations allowed by Bose statistics—boson sampling.

To make the problem more interesting, Aaronson allowed photons to be launched into any channel (as opposed to dropping all the ping-pong balls at the same spot), and they allowed each beam splitter to have adjustable phases (photons and phases are the key elements of an interferometer). By adjusting the locations of the input photon channels and the phases of the beam splitters, it would be possible to "program" this boson cascade to mimic interesting quantum systems or even to solve specific problems, although they were not thinking that far ahead. The main point of the paper was the proposal that implementing boson sampling in a photonic circuit uses resources that scale linearly in the number of photon channels while the problem solution grows exponentially—a clear quantum computational advantage.

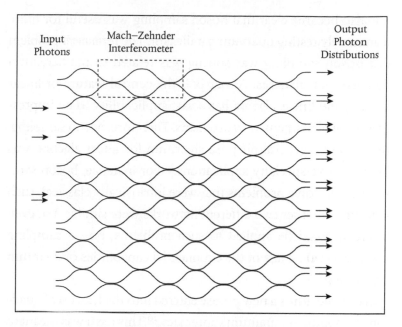

Figure 10.9 Boson-sampling circuit constructed from an array of integrated Mach–Zehnder interferometers. The four-photon input is repeatedly split into a four-photon superposition in the 16 channels at the output.

Shortly after the publication of Aaronson's and Arkhipov's paper, there was a flurry of experimental papers demonstrating boson sampling in the laboratory.[18] These were modest demonstrations that used only a few photons cascading through only a few beam splitters, and their results were easily simulated and verified using classical computers. But they were a crucial step, because these experiments proved the feasibility of the boson-sampling approach to quantum computing. Furthermore, these experiments tested the methods and materials that would be needed to implement larger systems that could be scaled up to sizes far beyond what any classical computer could simulate.

It also became clear that boson sampling was useful for more than merely testing quantum parallelism. The *permanent* problem that boson sampling was solving was related to mathematical properties of matrices. A matrix is the central feature of linear algebra, and linear algebra has a vast applicability to an impressive range of mathematical problems. Therefore, boson sampling, though not a general-purpose algorithm for linear algebra, was likely to have specialty applications.[19] For instance, boson sampling could solve problems that were intrinsically classical, such as testing whether two different networks were similar. Yet, even before it could be applied to such problems, boson sampling became a critical test of the exponential capabilities of quantum computing.

In the mid 2010s a new phrase entered into the lexicon of quantum computing—quantum supremacy.[20] This term was meant to express the ability of a quantum algorithm to surpass any possible classical computation. Up to that time, no quantum hardware had ever accomplished anything that a classical computer could not do better. The advantages of quantum computing were implicitly assumed, and worked great on paper, but quantum circuitry lagged far behind quantum algorithms. Therefore, the quest for quantum supremacy became the next holy grail in the field of quantum computing. If an actual quantum computer could be built that surpassed any possible calculation of a classical computer, then the era of quantum computing would have dawned.

In 2019, a research group at Google, led by John Martinis (1958–) of UC Santa Barbara, constructed a 54-qubit quantum computer with superconducting circuits that performed quantum random number generation. By testing the outputs on smaller versions and comparing the results with classical computations, they estimated that it would take 10,000 years for a classical computer to perform the same task.[21] This demonstration was touted with great fanfare by the Google team as well as by news outlets around the world. It also launched an immediate reaction from an IBM team who claimed that improved classical algorithms could perform the task in a matter of days. This generated a controversy about what constituted quantum supremacy and whether the Google demonstration had actually achieved it. It also raised objections to the phrase *quantum supremacy* itself because of unrelated but unpalatable uses of the second word. Therefore, the field distanced itself from that language and instead settled on the phrase *quantum computational advantage*. Regardless of the semantics, the Google-IBM controversy kept the door open for a more dramatic demonstration, and boson sampling stepped boldly through.

In 2020, a research team at the University of Science and Technology in China (USTC), led by Jian-Wei Pan (1970–), made a breakthrough for both boson sampling and photonic quantum computing. They constructed a photonic boson sampler using 50 states in 100 channels, publishing their results in *Science* with the title "Quantum Computational Advantage with Photons." This time, the claim stuck. The performance of the system exceeded any possible classical computation. A year later they expanded their system to 133 photons on 144 modes, further extending the computational advantage of photonic quantum computing.[22] The computer operated mostly at room temperature, which is an advantage over the cold temperatures needed for superconducting quantum computing. The superconducting elements in the single-photon detectors did require cooling, but the detectors can be coupled to the quantum hardware through fiber optics, making them external resources that can be cooled easily and cheaply.

Despite the impressive demonstration of quantum computational advantage by the Pan group at USTC, the free-space optical approach was not easily scalable. Sometimes the very Herculean effort needed to succeed at a piece of work also painfully highlights just how difficult that approach is. Therefore, more easily scalable technology will be needed for next-generation photonics quantum computers. To fill this need, PICs, with waveguides and beam splitters integrated directly on chips, are ideally suited to solve the scalability and integration problems of optical quantum computing.

A first step along this path was demonstrated with a general programmable room-temperature photonic quantum computer. In 2021, researchers at the Canadian company Xanadu, located in Toronto, published the experimental results of boson sampling on a special-purpose photonic chip[23] that performed boson sampling of strongly non-classical light. This was the first generally programmable photonic quantum computing chip, programmed using a quantum programming language they developed called Strawberry Fields. By simply changing the quantum coding (using a conventional computer interface), they switched the quantum computer output among three different quantum applications: transitions among states (spectra of molecular states), quantum docking, and similarity between graphs that represent two different molecules. These are radically different physics and math problems, yet the single chip could be reprogrammed using software to solve each one.

Such current demonstrations of photonic quantum computing are still merely toys compared with what they need to become in order to create markets willing to pay the kinds of money that the first systems will cost. But Moore's law of integrated circuits has every chance of being transferred to PICs. The first electronic integrated circuits in the 1960s look like jokes next to the Intel cores of today hosting billions of transistors with select physical features approaching the size of several atoms. Could anyone in 1970 have foreseen where that technology would go? The same is true for PICs. Photon losses in fabricated waveguides today are still limiting, and the degree of integration is modest, but there is a lot of room for improvement. The most important feature of PICs, what gives them such a bright future, is their intrinsic compatibility with interferometry. The nature of their small sizes and solid-state fabrication make them absolutely stable platforms. Instability, caused by shifting phases on unstable platforms, was always the biggest problem faced by interferometers, but this vanishes on the chip. Therefore, it is not hard to imagine a future, not too far off, when the sensitivity and versatility of optical interferometry, both classical and quantum, will permeate all the information devices that will envelope us.

In Defense of Optimism

My book *Mind at Light Speed*, published in 2001, described the optical revolution that was underway at that time, and I made a case in defense of optimism for the future of photonic science and technology. Optical fibers were just beginning to reach to the home, but it was not clear whether there would be a sustainable business model that would support widespread fiber distribution. Nonetheless, I maintained that optical technology is "easy" in the sense that one can almost always find a technical solution to an optics problem. There are few fundamental constraints on what can be done with lasers and lenses and fibers and materials, and many of the problems that are faced by new optical functionality tend to be technical rather than scientific. Clever designs by clever scientists and engineers usually overcome whatever the current limitations are. Therefore, back in 2001, although I had no crystal ball, I was confident that the advantages of fiber-optic communication would allow it to integrate seamlessly into daily life—eventually. Now fiber is back in full force, and even little towns in the American heartland are getting their fiber links. It would have been hard to predict in 2001 just how dominant streaming services have become today.

I can apply that same optimism to interference effects that lie at the heart of so many optical technologies. Interferometry is arguably the most sensitive measurement process that has ever been developed by mankind. We saw in Chapter 6 that interferometry was instrumental in measuring the first gravitational waves from black holes that merged far away across the universe. And interferometry using the full face of the Earth allowed us to make images of black holes in distant galaxies, as well as the one at the heart of our own galaxy. In the near future, there will be interferometric satellite systems that have baselines hundreds or thousands of times larger than the size of the Earth to detect ever more distant and smaller astronomical objects and ever weaker gravitational waves.

Turning from the astronomically large to the astronomically small, quantum information systems using interference and interferometry coupled to single atoms will perform quantum calculations to solve problems that could never be solved by classical computers. Photonic quantum computers, running special-purpose quantum software, will be positioned to optimize chemical synthesis, develop new pharmaceuticals, solve networking problems, and design new materials by using quantum simulation and quantum machine learning. Although there may be no need for quantum laptops any time soon, one never knows where market demands may go.

In many ways, optical technology today is similar to where electronic technology was in the last decades of the twentieth century. Much of current optical technology is crude and cumbersome and large and expensive, but it is changing rapidly. For instance, optical imaging technology has undergone a revolution with the incorporation of tiny studio-quality cameras in smart phones. The economies of scale in that market have generated sophisticated imaging capabilities at amazingly low cost. These same smartphone cameras could be used to detect holographic interference fringes for many types of applications, from high-precision laser ranging for home hobbyists, to microscopy of bacteria in medical samples for doctors working in resource-poor settings. In addition, holographic cameras and holographic displays are coming that will let people interact with their environments through augmented reality and interact with others either in virtual games or in virtual business settings.

The upshot is that there is plenty of time ahead, and there is plenty of room for new ideas on how to expand the uses of light and photons and interferometry and holography. Photonic technology one hundred years from now would seem like science fiction to us, but to those future generations, it will be as familiar and commonplace as opening one's eyes.

Notes

1

1. T. Young, "The Bakerian Lecture: Experiments and Calculations Relative to Physical Optics," *Philosophical Transactions of the Royal Society of London* 94 (1804): 1–16.
2. N. Burleigh, *Mirage: Napoleon's Scientists and the Unveiling of Egypt*, 1st ed. (New York: Harper, 2007).
3. Burleigh, *Mirage*, 126.
4. Malus recorded his activities on scientific expeditions to various parts of Egypt in his diary, published posthumously as *L'agenda de Malus. Souvenirs de l'expédition d'Égypte 1798–1801* (Paris, 1892).
5. Burleigh, *Mirage*, 213.
6. A. Robinson, *The Last Man Who Knew Everything: Thomas Young, the Anonymous Polymath Who Proved Newton Wrong, Explained How We See, Cured the Sick, and Deciphered the Rosetta Stone, Among Other Feats of Genius* (New York: Pi Press, 2006), 156.
7. "Young, Thomas." *Complete Dictionary of Scientific Biography*, vol. 14 (Charles Scribner's Sons, 2008), 562–572. *Gale Virtual Reference Library*.
8. O. Darrigol, *A History of Optics from Greek Antiquity to the Nineteenth Century* (Oxford: Oxford University Press, 2012), 152.
9. L. Euler, *Nova theoria lucis et colorum* (1746).
10. R. W. Home, "Euler's Anti-Newtonian Theory of Light," *Annals of Science* 45 (1988): 521–533.
11. The Bakerian Lectures were established in 1775 by Henry Baker to support a lecture by a fellow of the Royal Society on important topics in physical science.
12. J. Dornberg, "Count Rumford: The Most Successful Yank Abroad, Ever," *Smithsonian* 25 (1994): 102.
13. B. Thompson, "An Inquiry Concerning the Source of the Heat which Is Excited by Friction," *Philosophical Transactions of the Royal Society of London* 88 (1798): 80–102.

14. C. D. Andriesse, *Huygens: The Man Behind the Principle* (Cambridge: Cambridge University Press, 2010), 78.
15. J. G. Yoder, *Unrolling Time: Christiaan Huygens and the Mathematization of Nature* (Cambridge: Cambridge University Press, 1988), 89.
16. L. Kristjansson, "Iceland Spar and its Legacy in Science," *History of Geo- and Space Sciences* 3 (2012): 117–126.
17. Andriesse, *Huygens*, 266.
18. Andriesse, *Huygens*, 285.
19. Andriesse, *Huygens*, 287.
20. F. J. Dijksterhuis, *Lenses and Waves: Christiaan Huygens and the Mathematical Science of Optics in the Seventeenth Century* (Archimedes) (Boston; Dordrecht: Springer Netherlands, 2004), 167.
21. Andriesse, *Huygens*, 289.
22. Robinson, *The Last Man Who Knew Everything*, 108.
23. Robinson, *The Last Man Who Knew Everything*, 89.
24. T. Young, "The Bakerian Lecture: Experiments and Calculations Relative to Physical Optics," *Philosophical Transactions of the Royal Society of London* 94 (1804): 1–16.
25. Robinson, *The Last Man Who Knew Everything*, 112.
26. Robinson, *The Last Man Who Knew Everything*, 115.

2

1. J. F. W. Herschel, "Light," in *Encyclopaedia Metropolitana*, 4 ed. (London, 1845).
2. E. Malus, "Sur un propriété de la lumiere réfléchie," *Mém. Soc. Arc.* 2 (1809): 143–158. The Luxembourg anecdote is from Arago 1857, quoted in J. Z. Buchwald, *The Rise of the Wave Theory of Light: Optical Theory and Experiment in the Early Nineteenth Century* (Chicago: University of Chicago Press, 1989), 428.
3. Thomas Young, "Review of Laplace's memoir 'Sur la loi de la réfraction extraordinaire dans les cristaux diaphanes. Lu à la première Classe de l'Institut, dans sa séance du 30 janv. 1809. Journ. de physique, janv. 1809'," *Quarterly Rev.* 2 (1809): 337.
4. F. Arago, *Biographies of Distinguished Scientific Men*, 2d. Series (1859), 159.
5. T. Levitt, *A Short, Bright Light: Augustin Fresnel and the Birth of the Modern Lighthouse* (New York: W. W. Norton and Co., 2013), 25.
6. Levitt, *A Short, Bright Light*, 32.

7. J. Lequeux, *François Arago: A 19th Century French Humanist and Pioneer in Astrophysics* (Astrophysics and Space Science Library) (Springer International Publishing, 2015).

8. O. Darrigol, *A History of Optics from Greek Antiquity to the Nineteenth Century* (Oxford: Oxford University Press, 2012), 198.

9. Levitt, *A Short, Bright Light*, 40.

10. Levitt, *A Short, Bright Light*, 42.

11. There were arguably three theories of light at the time: waves, particles, and rays. Those who ascribed to each of these theories are called, respectively, undulationists, emissionists, and selectionists (a term proposed by historian of science Buchwald). By the early 1800s, the predominant theory was selectionist (rays), because it was more general by not needing to specify the nature of the light particles.

12. Levitt, *A Short, Bright Light*, 43.

13. B. Truesdell and L. Euler, *The Rational Mechanics of Flexible or Elastic Bodies, 1638–1788*. (Turici: O. Fussli, 1960).

14. D. Bernoulli, "Theoremata de oscillationibs corporum filo flexili connexorum et catenae verticaliter suspensae," *Comm. Acad. Sci. Petrop.* 6 (1740): 108–122. Submitted around 1732–1733.

15. D. Bernoulli, "Réflexions et éclaircissements sur les nouvelles vibrations des cordes," *Mem. Acad. Sci. Berlin* 9 (1755): 147–172. Submitted around 1753.

16. See chapter 4 of *Galileo Unbound* for the story of D'Alembert and Lagrange and Einstein.

17. D. Bernoulli (1755), 147–172.

18. M. Anderson, *Who Gave You the Epsilon?: And Other Tales of Mathematical History* (American Mathematical Society, 2009), 18.

19. B. Truesdell and L. Euler, 263.

20. A. C. Clairaut, *Table de la lune, calculées suivant la théorie de la gravitation universelle* (Paris, 1754).

21. See chapter 4 of *Galileo Unbound* for the tale of Maupertuis and the principle of least action.

22. J. Fourier, *Théorie Analytique de la Chaleur* (Paris: Firmin Didot Père et Fils, 1822).

23. J. Z. Buchwald, *The Rise of the Wave Theory of Light: Optical Theory and Experiment in the Early Nineteenth Century* (Chicago: University of Chicago Press, 1989), 156.

24. F. Arago, J. A. Barral, and P. Flourens, *Œuvres complètes de François Arago* (Paris: Gide et J. Baudry, 1854).

3

1. J. Lequeux, *François Arago: A 19th Century French Humanist and Pioneer in Astrophysics* (Astrophysics and Space Science Library) (Springer, 2015), v.
2. J. Michell, "On the means of discovering the distance, magnitude, &c. of the fixed stars in consequence of the diminution of the velocity of their light . . . By the Rev. John Michell . . . Read at the Royal Society, Nov. 27, 1783," (1783), 1.
3. Lequeux, *Arago*, 94.
4. He used an achromatic prism composed of two types of glass that deflects light but does not disperse it into a rainbow.
5. P. S. m. d. Laplace, *Exposition du système du monde* (English: The system of the world) (Paris, 1813).
6. F. Arago, *Œuvres*, tome 10, 312–333.
7. Lequeux, *Arago*, 263.
8. Humphrey Lloyd is most famous for demonstrating conical refraction that was predicted by his Irish compatriot William Rowan Hamilton.
9. Their daguerreotypes were not published by Fizeau and Foucault, but one was published by F. Arago and J. A. Barral, *Astronomie populaire* (Paris: Gide et J. Baudry, 1854).
10. J. Lequeux, *Hippolyte Fizeau physicien de la lumière* (Sciences et histoire) (Les Ulis: EDP Sciences, 2014).
11. W. Tobin, *The Life and Science of Léon Foucault: The Man who Proved the Earth Rotates* (Cambridge: Cambridge University Press, 2003).
12. Tobin, *The Life and Science of Léon Foucault*, 63. See also: *Ann. de Chim. et de Phys.*, 3-eme série, t. XXVI, 138 (1849).
13. Tobin, *The Life and Science of Léon Foucault*, 71. See also: Fizeau et Foucault, *Comptes rendus des séances de l'Académie des Science*, I. XXV (1847), 447.
14. L. Mandel and E. Wolf, "Coherence Properties of Optical Fields," *Reviews of Modern Physics* 37 (1965): 231
15. H. L. Fizeau, "Sur une expérience relative à la vitesse de propagation de la lumière," *Comptes rendus de l'Académie des sciences* 29 (1849): 90–92, 132.

16. Tobin, *The Life and Science of Léon Foucault*, 123.
17. Tobin, *The Life and Science of Léon Foucault*, 124.
18. Lequeux, *Hippolyte Fizeau*, 139.
19. Tobin, *The Life and Science of Léon Foucault*, 129.
20. H. Fizeau, "Acoustique et optique," presented at the *Société Philomathique de Paris* (Paris, 1848).
21. D. Nolte, "The Fall and Rise of the Doppler Effect," *Phys. Today* 73 (2020): 30.
22. J. Frercks, "Fizeau's Research Program on Ether Drag: A Long Quest for a Publishable Experiment," *Physics in Perspective* 7 (2005): 35–65.
23. H. Fizeau, "Sur les hypothèses relatives à l'éther lumineux," *Comptes Rendus de l'Acad. Sci., Paris* 33 (1851): 349–355.
24. J. Jamin and E. M. L. o. Bouty, *Cours de physique de l'École polytechnique*, 4. éd., augm. entiérement refondue ed. (Paris: Gauthier-Villars et fils, 1886).

4

1. A. A. Michelson, "Measurement by Light Waves," *Am. J. Sci.* 39, February Series 3 (1890): 115–121.
2. D. M. Livingston, *The Master of Light: A Biography of Albert A. Michelson* (New York: Charles Scribner's Sons, 1973), 119.
3. Livingston, *The Master of Light*, 27.
4. A. A. Michelson, "Experimental Determination of the Velocity of Light," *Proceedings of the American Association of Advanced Science* 27, (1878): 71–77.
5. R. Staley, *Einstein's Generation: The Origins of the Relativity Revolution* (Chicago: University of Chicago Press, 2008), 46.
6. Staley, *Einstein's Generation*, 49.
7. D. D. Nolte, "The Fall and Rise of the Doppler Effect," *Physics Today* 73 (2020): 31–35.
8. Livingston, *The Master of Light*, 88.
9. Livingston, *The Master of Light*, 90.
10. I have vivid memories of the beautiful parkway in the 1960s when my family would visit the Cleveland Art Museum situated on the north side of University Circle.
11. Livingston, *The Master of Light*, 3.

12. A. A. Michelson and E. W. Morley, "Influence of Motion of the Medium on the Velocity of Light," *Am. J. Sci.* 31 (1886): 377–386.
13. D. Nolte, "The Fall and Rise of the Doppler Effect," *Physics Today* 73 (2020): 31–35.
14. W. Voigt, "Ueber das Doppler'sche Princip," *Göttinger Nachrichten* 7 (1887): 41–51.
15. Livingston, *The Master of Light*, 122.
16. A. A. Michelson and E. W. Morley, "On the Relative Motion of the Earth and the Luminiferous Ether," *American Journal of Science* s3-34 (1887): 333.
17. For an excellent introduction to many of the different "named" interferometers, see Born and Wolf's very thorough *Principles of Optics* (Pergamon, 1980). For a more modern description of a wide array of interferometric approaches for measuring the optical properties of thin materials and molecular layers, see D. Nolte, *Optical Interferometry for Biology and Medicine* (Springer, 2011).
18. G. B. Airy, *Mathematical Tracts on the Lunar and Planetary Theories, the Figure of the Earth, Precession and Nutation, the Calculus of Variations, and the Undulatory Theory of Optics: Designed for the Use of Students in the University*, 2nd ed. (Cambridge, London, 1831).

5

1. D. H. De Vorkin, "Michelson and the Problem of Stellar Diameters," *Journal for the History of Astronomy* 6 (1975): 1–18.
2. F. W. Bessel, "Bestimmung der Entfernung des 61sten Sterns des Schwans" [English Title: Determination of the Distance to 61 Cygni]. *Astronomische Nachrichten* 16 (1838): 65–96.
3. H. Fizeau, "Sur une expérience relative à la vitesse de propagation de la lumière," *Comptes rendus de l'Académie des sciences* 29 (1849): 90–92, 132.
4. H. Fizeau, "Prix Bordin: Rapport sur le concours de l'annee 1867," *C. R. Acad. Sci.* 66 (1868): 932.
5. T. Henderson, "On the Parallax of Sirius," *Monthly Notices of the Royal Astronomical Society* 5 (1839): 5–7.
6. D. H. De Vorkin, "Michelson and the Problem of Stellar Diameters," *Journal for the History of Astronomy* 6 (1975): 1–18.

7. H. Stephan, *Comptes rendus,* lxxvi (1873): 1008–1010; lxxviii (1874): 1008–1012.

8. A. A. Michelson, "On the Application of Interference Methods to Astronomical Measurements," *The London, Edinburgh, and Dublin Philosophical Magazine and Journal of Science* 30 (1890): 1–21.

9. A. A. Michelson, "Measurement of Jupiter's Satellites by Interference," *Nature* 45 (1891): 160–161.

10. K. Schwarzschild, "Über messung von doppelsternen durch interferenzen," *Astron. Nachr.* 3335 (1896): 139.

11. https://galileo-unbound.blog/2019/09/15/karl-schwarzschilds-radius-how-fame-eclipsed-a-physicists-own-legacy/

12. For Betelgeuse L $\lambda/D = 1 \times 10^{-6}(5 \times 10^{18}/10^{12}) = 5$ m.

13. R Doradus in the southern constellation Dorado.

14. Livingston, *The Master of Light,* 239.

15. De Vorkin, "Michelson and the Problem of Stellar Diameters,": 11.

16. A. S. Eddington, "The Internal Constitution of the Stars," *Nature* cvi (1920): 14–20, 17; originally delivered as the opening address of Section A of the British Association, August 24, 1920, at Cardiff.

17. A. S. Eddington, *Monthly Notices of the Royal Astronomical Society* lxxxiii (1923): 309–316.

18. A. A. Michelson and F. G. Pease, "Measurement of the Diameter of α Orionis with the Interferometer," *Astrophysical Journal* 53 (1921): 249–259.

19. F. G. Pease, "Interferometer Methods in Astronomy," in *Ergebnisse der exakten naturwissenschaften* (Springer, 1931), 84–96.

20. A. Einstein, "Generation and Conversion of Light with Regard to a Heuristic Point of View," *Annalen der Physik* 17 (1905): 132–148.

21. G. I. Taylor, "Interference Fringes with Feeble Light," *Proceedings of the Cambridge Philosophical Society* 15 (1910): 114–115.

22. Hanbury Brown, *Boffin: A Personal Story of the Early Days of Radar, Radio Astronomy and Quantum Optics* (Bristol: Adam Hilger, 1991).

23. M. Ryle and D. D. Vonberg, "Solar Radiation on 175 Mc/sec," *Nature* 158 (1946): 339–340; K. I. Kellermann and J. M. Moran, "The Development of High-resolution Imaging in Radio Astronomy," *Annual Review of Astronomy and Astrophysics* 39 (2001): 457–509.

24. M. Ryle, "Solar Radio Emissions and Sunspots," *Nature* 161, no. 4082 (1948): 136.

25. R. H. Brown, *The Intensity Interferometer: Its Application to Astronomy* (London, New York: Taylor & Francis; Halsted Press, 1974).

26. Brown, *Boffin*, 106.
27. R. H. Brown and R. Q. Twiss, "A New Type of Interferometer for Use in Radio Astronomy," *Philosophical Magazine* 45, no. 366 (1954): 663–682.
28. R. H. Brown and R. Q. Twiss, "Correlation between Photons in 2 Coherent Beams of Light," *Nature* 177, no. 4497 (1956): 27–29.
29. I. Silva, and O. Freire, "The Concept of the Photon in Question: The Controversy Surrounding the HBT Effect circa 1956–1958," *Historical Studies in the Natural Sciences* 43, no. 4 (2013): 453–491.
30. A. Adam, L. Jánossy, and P. Varga, *Ann. der Phys.* 16 (1955): 408.
31. Brown, *Boffin*, 121.
32. E. M. Purcell, "Question of Correlation between Photons in Coherent Light Rays," *Nature* 178, no. 4548 (1956): 1448–1450.
33. R. H. Brown and R. Q. Twiss, "Test of a New Type of Stellar Interferometer on Sirius," *Nature* 178, no. 4541 (1956): 1046–1048.
34. H. W. Babcock, "The Possibility of Compensating Astronomical Seeing," *Publications of the Astronomical Society of the Pacific* 65, no. 386 (1953): 229–236.
35. J. C. Wyant, "White-light Extended Source Shearing Interferometer," *Applied Optics* 13, no. 1 (1974): 200–202.
36. R. Duffner and R. Fugate, *Adaptive Optics Revolution: A History* (Albuquerque: University of New Mexico Press, 2009), 46.
37. S. L. McCall, T. R. Brown, and A. Passner, "Improved Optical Stellar Image Using a Real-time Phase-correction System – Initial Results," *Astrophysical Journal* 211, no. 2 (1977): 463–468.
38. A. Buffington, F. S. Crawford, R. A. Muller, and C. D. Orth, "1st Observatory Results with an Image-sharpening Telescope," *Journal of the Optical Society of America* 67, no. 3 (1977): 304–305.
39. Duffner and Fugate, *Adaptive Optics Revolution*, 82.
40. A. Labeyrie, S. G. Lipson, and P. Nisenson, *An Introduction to Optical Stellar Interferometry* (Cambridge: Cambridge University Press, 2014), 186.
41. A. Labeyrie, "Interference Fringes Obtained on Vega with 2 Optical Telescopes," *Astrophysical Journal* 196, no. 2 (1975): L71–L75.
42. Duffner and Fugate, *Adaptive Optics Revolution*, 148.

6

1. Captain James Cook, *Captain Cook's Journal During His First Voyage Round the World*, Chapter III "Tahiti."

2. https://en.wikipedia.org/wiki/List_of_potentially_habitable_exoplanets

3. www.space.com/11877-alien-planets-search-canceled-missions-marcy.html

4. S. Lacour and Gravity Collaboration, "First Direct Detection of an Exoplanet by Optical Interferometry Astrometry and K-band Spectroscopy of HR 8799 e," *Astronomy & Astrophysics* 623 (2019): L11.

5. R. P. Butler and G. W. Marcy, "A Planet Orbiting 47 Ursae Majoris," *Astrophysical Journal* 464, no. 2 (1996): L153–L156.

6. R. P. Butler, G. W. Marcy, D. A. Fischer, T. M. Brown, A. R. Contos, S. G. Korzennik, P. Nisenson, and R. W. Noyes, "Evidence for Multiple Companions to Upsilon Andromedae," *Astrophysical Journal* 526, no. 2 (1999): 916–927.

7. K. I. Kellermann and J. M. Moran, "The Development of High-resolution Imaging in Radio Astronomy," *Annual Review of Astronomy and Astrophysics* 39 (2001): 457–509.

8. Ibid.

9. https://spectrum.ieee.org/aerospace/astrophysics/the-inside-story-of-the-first-picture-of-a-black-hole

10. www.extremetech.com/extreme/289423-it-took-half-a-ton-of-hard-drives-to-store-eht-black-hole-image-data

11. www.spectrum.ieee.org/aerospace/astrophysics/the-inside-story-of-the-first-picture-of-a-black-hole

12. M. Bartusiak, *Einstein's Unfinished Symphony: The Story of a Gamble, Two Black Holes, and a New Age of Astronomy* (New Haven: Yale University Press, 2017).

13. A. Einstein, "Näherungsweise Integration der Feldgleichungen der Gravitation," *Sitzungsberichte der Königlich Preussischen Akademie der Wissenschaften Berlin* 1 (1916): 688–696.

14. J. Weber, "Evidence for Discovery of Gravitational Radiation," *Physical Review Letters* 22, no. 24 (1969): 1320.

15. J. L. Cervantes-Cota, S. Galindo-Uribarri, and G. F. Smoot, "A Brief History of Gravitational Waves," *Universe* 2, no. 3 (2016): 22.

16. https://dcc.ligo.org/public/0038/P720002/001/P720002-00.pdf

17. Cervantes-Cota, "A Brief History of Gravitational Waves," 22.

18. Drever, Radio Interview, Archives, California Institute of Technology (1997). https://oralhistories.library.caltech.edu/272/1/Drever%2C%20R._OHO.pdf

19. J. Levin, *Black Hole Blues: And Other Songs from Outer Space* (New York: Anchor Books, 2017).
20. Levin, *Black Hole Blues*, 168.
21. B. C. Barish, "Nobel Lecture: LIGO and Gravitational Waves II," *Reviews of Modern Physics* 90, no. 4 (2018): Art no. 040502, 6.
22. Barry Barish clarifies the distinction between an engineering run and an observational run: "For LIGO, (the engineering run) represents the last chance for sub-system leaders or operational leaders to 'freeze' the crucial running parameters for the interferometer. We try very hard not to change much after we begin a data run, because we do an 'out of time' analysis to determine the statistical significance and that took a month for the first event. Obviously, that needs to be done for essentially the same running conditions. So, an engineering run for us is the time to freeze running conditions, while for most complicated instruments, it has more of a test function. This distinction is why we could use engineering run data to observe the event, but needed ~ one month of off coincidence time data to determine the significance." Personal communication.
23. Personal communication from Barry Barish.
24. www.sciencemag.org/news/2016/02/we-did-it-voices-gravitational-wave-press-conference
25. It was after watching the press conference on the detection of gravitational waves in Feb. 2016 that I decided to write this book.
26. https://en.wikipedia.org/wiki/List_of_gravitational_wave_observations

7

1. Ernst Abbe, "Über die Grenzen der geometrischen Optik" [English title: Beyond the Limits of Geometric Optics], *Jenaische Zeitschrift für Naturwissenschaft* (Jena, Germany: Verlag Von Gustav Fischer, 1878): 71–109.
2. M. W. Jackson, *Spectrum of Belief: Joseph von Fraunhofer and the Craft of Precision Optics* (Cambridge, MA: MIT Press, 2000), 53.
3. www.fraunhofer.de/content/dam/zv/en/documents/Glashuette_engl_tcm6-106162_tcm63-778.pdf
4. Jackson, *Spectrum of Belief*, 59.

5. J. Fraunhofer, "Bestimmung des Brechungs- und des Farben-Zerstreuungs – Vermögens verschiedener Glasarten, in Bezug auf die Vervollkommnung achromatischer Fernröhre" [English title: Determination of the Refractive and Color-dispersing Power of Different Types of Glass, in Relation to the Improvement of Achromatic Telescopes], *Annalen der Physik* 56 (1817): 264–313.

6. J. Fraunhofer, "Kurzer bericht von den resultaten neuerer versuche uber die gesetze des lichtes and die theorie derselben," *Gilberts Ann.* 74 (1823): 337–378.

7. G. K. Sweetnam, *The Command of Light: Rowland's School of Physics and the Spectrum* (Memoirs of the American Philosophical Society held at Philadelphia for Promoting Useful Knowledge, 0065–9738; v. 238) (Philadelphia: American Philosophical Society, 2000).

8. Sweetnam, *The Command of Light*, xvii.

9. C. N. Brown, "The Ruling Engines and Diffraction Gratings of Henry Augustus Rowland," *Annals of Science* (2022).

10. Ibid., 125.

11. H. Kohler, "On Abbe's Theory of Image-formation in the Microscope," *Optica Acta* 28, no. 12 (1981): 1691–1701.

12. F. Zernike, "How I Discovered Phase Contrast," *Science* 121, no. 3141 (1955): 345–349.

13. Ibid.

14. Ibid.

15. This and following comments in this section from personal communication with Prof. John Sedat on February 2, 2022.

16. S. W. Hell, S. Lindek, C. Cremer, and E. H. K. Stelzer, "Measurement of the 4pi-confocal Point-spread Function Proves 75 nm Axial Resolution," *Applied Physics Letters* 64, no. 11 (1994): 1335–1337.

17. M. G. L. Gustafsson, D. A. Agard, and J. W. Sedat, "(IM)-M-5: 3D Widefield Light Microscopy with Better than 100 nm Axial Resolution," *Journal of Microscopy* 195 (1999): 10–16.

18. W. Lukosz, "Optical Systems with Resolving Powers Exceeding Classical Limit," *Journal of the Optical Society of America* 56, no. 11 (1966): 1463.

19. M. G. L. Gustafsson, "Surpassing the Lateral Resolution Limit by a Factor of Two Using Structured Illumination Microscopy," *Journal of Microscopy* 198 (2000): 82–87.

20. L. Shao, B. Isaac, S. Uzawa, D. A. Agard, J. W. Sedat, and M. G. L. Gustafsson, "(IS)-S-5: Wide-field Light Microscopy with 100-nm-scale

Resolution in Three Dimensions," *Biophysical Journal* 94, no. 12 (2008): 4971–4983.
21. www.nobelprize.org/prizes/chemistry/2014/hell/biographical/

8

1. S. F. Johnston, "From White Elephant to Nobel Prize: Dennis Gabor's Wavefront Reconstruction," *Historical Studies in the Physical and Biological Sciences*, Review 36 (2005): 35–70.
2. Letter from Bragg to Gabor July 4, 1948 quoted in S. F. Johnston, *Holographic Visions: A History of New Science* (Oxford: Oxford University Press, 2006).
3. S. Johnston, *Holographic Visions: A History of New Science* (Oxford: Oxford University Press, 2016), 16.
4. Johnston, *Holographic Visions*, 90.
5. E. N. Leith and J. Upatnieks, "New Techniques in Wavefront Reconstruction," *Journal of the Optical Society of America* 51 (1961): 1469.
6. E. N. Leith and J. Upatnieks, "Reconstructed Wavefronts and Communication Theory," *Journal of the Optical Society of America* 52, no. 10 (1962): 1123.
7. J. Hecht, *Beam: The Race to Make the Laser* (Oxford: Oxford University Press, 2005); C. H. Townes, *How the Laser Happened* (New York: Oxford University Press, 1999).
8. W. E. Lamb Jr. and R. C. Retherford, "Fine Structure of the Hydrogen Atom, Part I," *Phys. Rev.* 79 (1950): 549–572.
9. E. M. Purcell and R. V. Pound, "A Nuclear Spin System at Negative Temperature," *Phys. Rev.* 81 (1951): 279–280.
10. A. L. Schawlow and C. H. Townes, "Infrared and Optical Masers," *Phys. Rev.* 112 (1958): 1940–1949.
11. Hecht, *Beam*, 141.
12. Johnston, *Holographic Visions*, 111.
13. https://americanhistory.si.edu/collections/search/object/nmah_1448340
14. Johnston, *Holographic Visions*, 112.
15. Johnston, *Holographic Visions*, 113.
16. Novotny quoted in Johnston, *Holographic Visions*, 118.
17. Stephen A. Benton "Holography Reinvented," *Proc. SPIE 4737, Holography: A Tribute to Yuri Denisyuk and Emmett Leith* (2002), https://doi.org/10.1117/12.474954

18. P. J. Vanheerden, "Theory of Optical Information Storage in Solids," *Applied Optics* 2, no. 4 (1963): 393–400.

19. Hecht, *Beam*, p. 162.

20. A. Ashkin, G. D. Boyd, J. M. Dziedzic, R. G. Smith, A. A. Ballman, J. J. Levinstein, and K. Nassau, "Optically-induced Refractive Index Inhomogeneities in $LiNbO_3$ and $LiTaO_3$ (Ferroelectric materials - Nonlinear optics - E," *Applied Physics Letters* 9, no. 1 (1966): 72.

21. F. S. Chen, J. T. Lamacchia, and D. B. Fraser, "Holographic Storage in Lithium Niobate," *Applied Physics Letters* 13, no. 7 (1968): 223.

22. Private communication with Alastair Glass, 2022.

23. A. M. Glass, D. V. D. Linde, and T. J. Negran, "High-voltage Bulk Photovoltaic Effect and the Photorefractive Process in $LiNbO_3$," *Applied Physics Letters* 25, no. 4 (1974): 233–235.

24. I. Lahiri, L. J. Pyrak-Nolte, D. D. Nolte, M. R. Melloch, R. A. Kruger, G. D. Bacher, and M. B. Klein, "Laser-based Ultrasound Detection using Photorefractive Quantum Wells," *Appl. Phys. Lett.* 73 (1998): 1041–1043.

25. D. Psaltis, D. Brady, X. G. Gu, and S. Lin, "Holography in Artificial Neural Networks," *Nature* 343, no. 6256 (1990): 325–330; G. Zhou and D. Z. Anderson, "Acoustic-signal Recognition with a Photorefractive Time-delay Neural-network," *Optics Letters* 19 (1994): 655–7.

26. Z. Li, H. Sun, J. Turek, S. Jalal, M. Childress, and D. D. Nolte, "Doppler Fluctuation Spectroscopy of Intracellular Dynamics in Living Tissue," *Journal of the Optical Society of America A: Optics, Image Science, and Vision* 36, no. 4 (2019): 665–677.

27. Z. Li, R. An, W. M. Swetzig, M. Kanis, N. Nwani, J. Turek, D. Matei, and D. Nolte, "Intracellular Optical Doppler Phenotypes of Chemosensitivity in Human Epithelial Ovarian Cancer," *Scientific Reports* 10, no. 1 (2020): 17354.

28. K. Jeong, J. J. Turek, and D. D. Nolte, "Imaging Motility Contrast in Digital Holography of Tissue Response to Cytoskeletal Anti-cancer Drugs," *Optics Express* 15 (2007): 14057–14064.

29. Y. Park, C. Depeursinge, and G. Popescu, "Quantitative Phase Imaging in Biomedicine," *Nature Photonics* 12, no. 10 (2018): 578–589.

9

1. L. Mandel, "Quantum Theory of Interference Effects Produced by Independent Light Beams," *Physical Review* 134, no. 1A (1964): A10;

P. A. M. Dirac, *Quantum Mechanics*, 4th ed. (Oxford: Clarendon Press, 1958): 9.

2. In a December 1926 letter to Max Born, Einstein wrote: "The theory produces a good deal but hardly brings us closer to the secret of the Old One. I am at all events convinced that He does not play dice."

3. A. Einstein, "Generation and Conversion of Light with Regard to a Heuristic Point of View," *Annalen Der Physik* 17, no. 6 (1905):132–148. Gilbert Lewis named the photon in 1926.

4. A. Whitaker, *Einstein, Bohr and the Quantum Dilemma* (Cambridge: Cambridge University Press, 1996).

5. Letter from Einstein to Max Born, March 3, 1947; *The Born–Einstein Letters: Correspondence between Albert Einstein and Max and Hedwig Born from 1916 to 1955* (New York: Walker, 1971).

6. The EPR paper does not mention "paradox." This word was applied later by Schrödinger commenting on the EPR paper. It was also Schrödinger who coined the term entanglement, which was the motivation for his absurd cat.

7. A. Einstein, B. Podolsky, and N. Rosen, "Can Quantum-mechanical Description of Physical Reality Be Considered Complete?" *Physical Review* 47, no. 10 (1935): 0777–0780.

8. N. Harrigan and R. W. Spekkens, "Einstein, Incompleteness, and the Epistemic View of Quantum States," *Foundations of Physics* 40, no. 2 (2010): 125–157.

9. E. Schrödinger, "Discussion of Probability Relations between Separated Systems," *Proceedings of the Cambridge Philosophical Society* 31 (1935): 555–563.

10. F. D. Peat, *Infinite Potential: The Life and Times of David Bohm* (Reading, MA: Helix Books, Addison-Wesley Publishing Company, Inc., 1996).

11. D. Bohm, "A Suggested Interpretation of the Quantum Theory in Terms of Hidden Variables .1," *Physical Review* 85, no. 2 (1952): 166–179; D. Bohm, "A Suggested Interpretation of the Quantum Theory in Terms of Hidden Variables .2," *Physical Review* 85, no. 2 (1952): 180–193.

12. Y. Aharonov and D. Bohm, "Significance of Electromagnetic Potentials in the Quantum Theory," *Physical Review* 115, no. 3 (1959): 485–491.

13. D. Bohm and Y. Aharonov, "Discussion of Experimental Proof for the Paradox of Einstein Rosen and Podolsky," *Physical Review* 108, no. 4 (1957): 1070–1076.

14. C. S. Wu, *Phys. Rev.* 77 (1950): 136.

15. Letter from Einstein to Max Born, March 3, 1947; *The Born–Einstein Letters: Correspondence between Albert Einstein and Max and Hedwig Born from 1916 to 1955* (New York: Walker, 1971).

16. J. S. Bell, R. A. Bertlmann, and A. Zeilinger, *Quantum [Un]speakables: From Bell to Quantum Information* (Berlin: Springer, 2002).

17. Quote by Paul Ehrenfest from Whitaker, *Einstein, Bohr and the Quantum Dilemma*, 210.

18. J. Bernstein, *Quantum Profiles* (Princeton: Princeton University Press, 2001).

19. J. von Neumann, *Mathematical Foundations of Quantum Mechanics*, 1996 ed. (Princeton: Princeton University Press, 1932).

20. J. Bell, "On the Einstein-Podolsky-Rosen Paradox," *Physics* 1 (1964): 195.

21. John Clauser APS Oral Histories interview, www.aip.org/history-programs/niels-bohr-library/oral-histories/25096</IBT>

22. Ibid.

23. Ibid.

24. Ibid.

25. S. J. Freedman and J. F. Clauser, "Experimental Test of Local Hidden-variable Theories," *Physical Review Letters* 28, no. 14 (1972): 938.

26. A. Aspect, P. Grangier, and G. Roger, "Experimental Realization of Einstein-Podolsky-Rosen-Bohm *gedankenexperiment* – a New Violation of Bell Inequalities," *Physical Review Letters* 49, no. 2 (1982): 91–94.

27. M. Born and E. Wolf, *Principles of Optics: Electromagnetic Theory of Propagation, Interference and Diffraction of Light* (Cambridge: Cambridge University Press, 1969).

28. R. L. Pfleegor and L. Mandel, "Interference Effects at Single Photon Level," *Physics Letters A* A 24, no. 13 (1967): 766.

29. P. A. M. Dirac, *Quantum Mechanics*, 4th ed. (Oxford: Clarendon Press, 1958), 9.

30. Down-conversion is a solid-state analog of *resonance fluorescence* in atomic optics that had been used by Mandel and his student Jeff Kimble in 1977 to demonstrate photon anti-bunching (H. J. Kimble, M. Dagenais, and L. Mandel, "Photon Anti-bunching in Resonance Fluorescence," *Physical Review Letters* 39, no. 11 (1977): 691–695). This is the opposite effect of normal light sources which show photon bunching and which Purcell had invoked to explain the quantum origin of the HBT correlation results. Photon anti-bunching is a quantum form of light that has no classical analog, and with the

advent of parametric down-conversion techniques, true two-photon light sources were now available.

31. R. Ghosh, C. K. Hong, Z. Y. Ou, and L. Mandel, "Interference of 2 Photons in Parametric Down Conversion," *Physical Review A* 34, no. 5 (1986): 3962–3968.

32. The absence of amplitude interference is because the signal and idler beams have a wide bandwidth and the single-photon interference fringes wash out. But the two-photon interference fringes are independent of the bandwidth.

33. M. A. Horne, A. Shimony, and A. Zeilinger, "2-particle Interferometry," *Physical Review Letters* 62, no. 19 (1989): 2209–2212.

34. P. G. Kwiat, W. A. Vareka, C. K. Hong, H. Nathel, and R. Y. Chiao, "Correlated 2-photon Interference in a Dual-beam Michelson Interferometer," *Physical Review A* 41, no. 5 (1990): 2910–2913; Z. Y. Ou, X. Y. Zou, L. J. Wang, and L. Mandel, "Observation of Nonlocal Interference in Separated Photon Channels," *Physical Review Letters* 65, no. 3 (1990): 321–324; J. G. Rarity, P. R. Tapster, E. Jakeman, T. Larchuk, R. A. Campos, M. C. Teich, and B. E. A. Saleh, "2-photon Interference in a Mach–Zehnder Interferometer," *Physical Review Letters* 65, no. 11 (1990): 1348–1351.

35. Z. Y. Ou, X. Y. Zou, L. J. Wang, and L. Mandel, "Observation of Nonlocal Interference in Separated Photon Channels," *Physical Review Letters* 65, no. 3 (1990): 321–324.

36. J. D. Franson, "Bell Inequality for Position and Time," *Physical Review Letters* 62, no. 19 (1989): 2205–2208.

37. L. Mandel and E. Wolf, *Optical Coherence and Quantum Optics* (Cambridge: Cambridge University Press, 1995).

38. www.youtube.com/watch?v=ooZvkPgy7-Y

39. C. H. Bennett and G. Brassard, "Quantum Cryptography: Public Key Distribution and Coin Tossing," *Theoretical Computer Science* 560 (2014): 7–11.

40. A. Peres and W. K. Wootters, "Optimal Detection of Quantum Information," *Physical Review Letters* 66, no. 9 (1991): 1119–1122.

41. W. K. Wootters and W. H. Zurek, "A Single Quantum Cannot Be Cloned," *Nature* 299, no. 5886 (1982): 802–803.

42. www.youtube.com/watch?v=ooZvkPgy7-Y

43. D. Bouwmeester, J.-W. Pan, K. Mattle, M. Eibl, H. Weinfurter, and A. Zeilinger, "Experimental Quantum Teleportation," *Nature* 390 (1997): 575–579.

44. D. Boschi, S. Branca, F. De Martini, L. Hardy, and S. Popescu, "Experimental Realization of Teleporting an Unknown Pure Quantum State via Dual Classical and Einstein-Podolsky-Rosen Channels," *Phys. Rev. Lett.* 80 (1998): 1121–1125.

45. www.youtube.com/watch?v=JY6UCVsiTPU

46. D. M. Greenberger, M. A. Horne, and A. Zeilinger, "Going Beyond Bell Theorem," in *1988 Fall Workshop on Bells Theorem, Quantum Theory and Conceptions of the Universe*, George Mason Univ., Fairfax, VA, Oct. 21–22, 1988, in *Fundamental Theories of Physics* 37 (1989): 69–72.

47. R. Ursin, T. Jennewein, M. Aspelmeyer, R. Kaltenbaek, M. Lindenthal, P. Walther, and A. Zeilinger, "Communications – Quantum Teleportation across the Danube," *Nature* 430, no. 7002 (2004): 849.

48. J. Yin, J. G. Ren, H. Lu, Y. Cao, H. L. Yong, Y. P. Wu, C. Liu, S. K. Liao, F. Zhou, Y. Jiang, X. D. Cai, P. Xu, G. S. Pan, J. J. Jia, Y. M. Huang, H. Yin, J. Y. Wang, Y. A. Chen, C. Z. Peng, and J. W. Pan, "Quantum Teleportation and Entanglement Distribution over 100-kilometre Free-space Channels," *Nature* 488, no. 7410 (2012): 185–188.

49. X. S. Ma, T. Herbst, T. Scheidl, D. Q. Wang, S. Kropatschek, W. Naylor, B. Wittmann, A. Mech, J. Kofler, E. Anisimova, V. Makarov, T. Jennewein, R. Ursin, and A. Zeilinger, "Quantum Teleportation over 143 Kilometres Using Active Feed-forward," *Nature* 489, no. 7415 (2012): 269–273.

50. J. G. Ren, P. Xu, H. L. Yong, L. Zhang, S. K. Liao, J. Yin, W. Y. Liu, W. Q. Cai, M. Yang, L. Li, K. X. Yang, X. Han, Y. Q. Yao, J. Li, H. Y. Wu, S. Wan, L. Liu, D. Q. Liu, Y. W. Kuang, Z. P. He, P. Shang, C. Guo, R. H. Zheng, K. Tian, Z. C. Zhu, N. L. Liu, C. Y. Lu, R. Shu, Y. A. Chen, C. Z. Peng, J. Y. Wang, and J. W. Pan, "Ground-to-satellite Quantum Teleportation," *Nature* 549, no. 7670 (2017): 70.

51. D. Llewellyn, Y. H. Ding, Faruque, II, S. Paesani, D. Bacco, R. Santagati, Y. J. Qian, Y. Li, Y. F. Xiao, M. Huber, M. Malik, G. F. Sinclair, X. Q. Zhou, K. Rottwitt, J. L. O'Brien, J. G. Rarity, Q. H. Gong, L. K. Oxenlowe, J. W. Wang, and M. G. Thompson, "Chip-to-chip Quantum Teleportation and Multi-photon Entanglement in Silicon," *Nature Physics* 16, no. 2 (2020): 148.

52. B. J. Metcalf, J. B. Spring, P. C. Humphreys, N. Thomas-Peter, M. Barbieri, W. S. Kolthammer, X. M. Jin, N. K. Langford, D. Kundys, J. C. Gates, B. J. Smith, P. G. R. Smith, and I. A. Walmsley, "Quantum Teleportation on a Photonic Chip," *Nature Photonics* 8, no. 10 (2014): 770–774.

53. A. Furusawa et al., "Unconditional Quantum Teleportation," *Science* 282 (1998): 706–709.

54. N. Takei et al., "Experimental Demonstration of Quantum Teleportation of a Squeezed State," *Phys. Rev. A* 72 (2005): 042304.

55. N. Lee et al., "Teleportation of Nonclassical Wave Packets of Light," *Science* 332 (2011): 330–333.

10

1. D. Deutsch, "Quantum-theory, the Church–Turing Principle and the Universal Quantum Computer," *Proceedings of the Royal Society of London Series A: Mathematical, Physical, and Engineering Sciences* 400, no. 1818 (1985): 97–117.

2. https://nautil.us/issue/66/clockwork/haunted-by-his-brother-he-revolutionized-physics-rp

3. C. W. Misner, K. S. Thorne, and J. A. Wheeler, *Gravitation* (New York: W. H. Freeman and Company, 1973).

4. www.scientificamerican.com/article/hugh-everett-biography/

5. O. Morton, "The Computable Cosmos of David Deutsch," *American Scholar* 69, no. 3 (2000): 51–67.

6. Ibid., 58.

7. D. Deutsch, "Quantum-theory, the Church–Turing Principle and the Universal Quantum Computer," *Proceedings of the Royal Society of London Series A: Mathematical, Physical, and Engineering Sciences* 400, no. 1818 (1985): 97–117.

8. D. Deutsch and R. Jozsa, "Rapid Solution of Problems by Quantum Computation," *Proceedings of the Royal Society of London Series A: Mathematical, Physical, and Engineering Sciences* 439, no. 1907 (1992): 553–558.

9. www.forbes.com/sites/johnkoetsier/2021/04/07/photonic-supercomputer-for-ai-10x-faster-90-less-energy-plus-runway-for-100x-speed-boost

10. R. L. Rivest, A. Shamir, and L. M. Adleman, "A Method of Obtaining Digital Signatures and Public-key Cryptosystems," *Commun. ACM* 21, no. 2 (1978): 120–126.

11. Interview with *Nature* magazine, Oct. 30, 2020. www.nature.com/articles/d41586-020-03068-9

12. P. W. Shor, "Algorithms for Quantum Computation – Discrete Logarithms and Factoring," in *35th Annual Symposium on Foundations of*

Computer Science, Proceedings, S. Goldwasser ed. (Annual Symposium on Foundations of Computer Science, 1994), 124–134.

13. There are other equivalent unitary matrices for beam splitters that have different overall phases, but the Hadamard is a particularly simple choice.

14. E. Knill, R. Laflamme, and G. J. Milburn, "A Scheme for Efficient Quantum Computation with Linear Optics," *Nature* 409, no. 6816 (2001): 46–52.

15. Three months after KLM submitted their paper, a group from Munich Germany proposed a more direct form of measurement-based linear optical quantum computing that introduced "cluster states" of many highly entangled qubits. See R. Raussendorf and H. J. Briegel, "A One-way Quantum Computer," *Physical Review Letters* 86, no. 22 (2001): 5188–5191.

16. J. Carolan, C. Harrold, C. Sparrow, E. Martin-Lopez, N. J. Russell, J. W. Silverstone, P. J. Shadbolt, N. Matsuda, M. Oguma, M. Itoh, G. D. Marshall, M. G. Thompson, J. C. F. Matthews, T. Hashimoto, J. L. O'Brien, and A. Laing, "Universal Linear Optics," *Science* 349, no. 6249 (2015): 711–716.

17. https://complexityzoo.net/Complexity_Zoo

18. J. B. Spring et al., "Photonic Boson Sampling in a Tunable Circuit," *Science* 339, no. 6121 (2013): 794–798.

19. T. R. Bromley, J. M. Arrazola, S. Jahangiri, J. Izaac, N. Quesada, A. D. Gran, M. Schuld, J. Swinarton, Z. Zabaneh, and N. Killoran, "Applications of Near-term Photonic Quantum Computers: Software and Algorithms," *Quantum Science and Technology* 5, no. 3 (2020): 034010.

20. A. W. Harrow and A. Montanaro, "Quantum Computational Supremacy," *Nature* 549, no. 7671 (2017): 203–209.

21. F. Arute, J. M. Martinis, et al., "Quantum Supremacy Using a Programmable Superconducting Processor," *Nature* 574, no. 7779 (2019): 505.

22. H. S. Zhong, J. W. Pan, et al., "Phase-programmable Gaussian Boson Sampling Using Stimulated Squeezed Light," *Physical Review Letters* 127, no. 18 (2021): 180502.

23. J. M. Arrazola et al., "Quantum Circuits with Many Photons on a Programmable Nanophotonic Chip," *Nature* 591, no. 7848 (2021): 54.

Bibliography

1. Thomas Young Polymath

Andriesse, C. D. *Huygens: The Man Behind the Principle*. Cambridge: Cambridge University Press, 2010.

Buchwald, J. Z. *The Rise of the Wave Theory of Light: Optical Theory and Experiment in the Early Nineteenth Century*. Chicago: University of Chicago Press, 1989.

Burleigh, N. *Mirage: Napoleon's Scientists and the Unveiling of Egypt*. New York: Harper, 2007.

Dijksterhuis, F. J. *Lenses and Waves: Christiaan Huygens and the Mathematical Science of Optics in the Seventeenth Century*. Dordrecht Netherlands: Springer, 2004.

Robinson, A. *The Last Man Who Knew Everything: Thomas Young, the Anonymous Polymath who Proved Newton Wrong, Explained How We See, Cured the Sick, and Deciphered the Rosetta Stone, Among Other Feats of Genius*. New York: Pi Press, 2006.

Yoder, J. G. *Unrolling Time: Christiaan Huygens and the Mathematization of Nature*. Cambridge: Cambridge University Press, 1988.

"Young, Thomas." *Complete Dictionary of Scientific Biography*, vol. 14. Charles Scribner's Sons, 562–572. Gale Virtual Reference Library, 2008.

2. The Fresnel Connection

Boutry, G. A. "Augustin Fresnel: His Time, Life and Work, 1788–1827." *Science Progress* 36 (1948): 587–604.

"Fresnel, Augustin Jean." *Complete Dictionary of Scientific Biography*, vol. 5, Charles Scribner's Sons, 165–171. Gale Virtual Reference Library, 2008.

Home, R. W. "Euler's Anti-Newtonian Theory of Light." *Annals of Science* 45, no. 5 (1988): 521–533.

Kipnis, N. S. *History of the Principle of Interference of Light*. Basel: Birkhäuser Verlag, 1991.

Lequeux, J. *François Arago A 19th Century French Humanist and Pioneer in Astrophysics*. Cham: Springer International Publishing, 2015.

Levitt, T. *A Short, Bright Light: Augustin Fresnel and the Birth of the Modern Lighthouse*. New York: W. W. Norton and Co., 2013.

Truesdell, B. and L. Euler. *The Rational Mechanics of Flexible or Elastic Bodies, 1638–1788*. Turici: O. Fussli, 1960.

3. At Light Speed

Connes, P. "From Newtonian Fits to Wellsian Heat Rays – The History of Multiple-beam Interference." *Journal of Optics-Nouvelle Revue d'Optique* 17 (1986): 5–28.

Darrigol, O. *A History of Optics from Greek Antiquity to the Nineteenth Century*. Oxford: Oxford University Press, 2012.

Frercks, J. "Fizeau's Research Program on Ether Drag: A Long Quest for a Publishable Experiment." *Physics in Perspective* 7, no. 1 (2005): 35–65.

Genzel, L. and K. Sakai. "Interferometry from 1950 to Present." *Journal of the Optical Society of America* 67 (1977): 871–879.

Harman, P. M. *Energy, Force, and Matter: The Conceptual Development of Nineteenth-Century Physics*. Cambridge: Cambridge University Press, 1982.

Lequeux, J. *Hippolyte Fizeau Physicien de la Lumière (Sciences et histoire)*. Les Ulis: EDP Sciences, 2014.

Livingston, D. M. *The Master of Light: A Biography of Albert A. Michelson*. New York: Scribner, 1973.

Pippard, B. "Dispersion in the Ether: Light over the Water." *Physics in Perspective* 3 (2001): 258–270.

Purrington, R. D. *Physics in the Nineteenth Century*. New Brunswick, N.J: Rutgers University Press, 1997.

Staley, R. *Einstein's Generation: The Origins of the Relativity Revolution*. Chicago: University of Chicago Press, 2008.

Tobin, W. *The Life and Science of Léon Foucault: The Man Who Proved the Earth Rotates*. Cambridge: Cambridge University Press, 2003.

Vaughan, J. M. "Interferometry, Atoms and Light Scattering: One Hundred Years of Optics." *Journal of Optics A: Pure and Applied Optics* 1 (1999): 750–768.

Whittaker, E. T. *A History of the Theories of Aether and Electricity*. New York: Tomash Publishers; American Institute of Physics, 1987.

4. After the Gold Rush

Anderson, R. H., R. Bilger, and G. E. Stedman. "Sagnac Effect – A Century of Earth-rotated Interferometers." *American Journal of Physics* 62 (1994): 975–985.

Bouchareine, P. "Charles Fabry Metrologist." *Annales De Physique* 21 (1996): 589–600.

Connes, P. "From Newtonian Fits to Wellsian Heat Rays – The History of Multiple-beam Interference." *Journal of Optics-Nouvelle Revue d'Optique* 17 (1986): 5–28.

Courtes, G. "Astrophysical Applications of Fabry–Perot Multiple Beam Interferometry by Fabry, his Colleagues and Followers." *Journal of Optics-Nouvelle Revue d'Optique* 17 (1986): 29–34.

Gangopadhyay, T. K. and P. J. Henderson. "Vibration: History and Measurement with an Extrinsic Fabry–Perot Sensor with Solid-state Laser Interferometry." *Applied Optics* 38 (1999): 2471–2477.

Mulligan, J. F. "Who Were Fabry and Perot?" *American Journal of Physics* 66 (1998): 797–802.

Quirrenbach, A. "The Development of Astronomical Interferometry." *Experimental Astronomy* 26 (2009): 49–63.

Schaffner, K. F. *Nineteenth-Century Aether Theories.* Oxford: Pergamon Press, 1972.

5. Stellar Interference

Brown, R. H. and R. Q. Twiss. "Correlation Between Photons in 2 Coherent Beams of Light." *Nature* 177 (1956): 27–29.

Brown, R. H. and R. Q. Twiss. "Test of a New Type of Stellar Interferometer on Sirius." *Nature* 178 (1956): 1046–1048.

Buffington, A., F. S. Crawford, S. M. Pollaine, C. D. Orth, and R. A. Muller. "Sharpening Stellar Images." *Science* 200 (1978): 489–494.

Duffner, R. and R. Fugate. *Adaptive Optics Revolution: A History.* Albuquerque: University of New Mexico Press, 2009.

Fugate, R. Q., D. L. Fried, G. A. Ameer, B. R. Boeke, S. L. Browne, P. H. Roberts, R. E. Ruane, G. A. Tyler, and L. M. Wopat. "Measurement of Atmospheric Wave-front Distortion Using Scattered-light from a Laser Guide-star," *Nature* 353 (1991): 144–146.

Glindemann, A. *Principles of Stellar Interferometry.* Berlin, Heidelberg: Springer, 2011.

Guyon, O. "Extreme Adaptive Optics." In *Annual Review of Astronomy and Astrophysics* 56, edited by S. M. Faber and E. VanDishoeck, 315–355. San Mateo, CA: Annual Reviews, 2018.

Hardy, J. W. "Adaptive Optics." *Scientific American* 270, no. 6 (1994): 60–65.

Hearnshaw, J. B. *The Analysis of Starlight: Two Centuries of Astronomical Spectroscopy.* Cambridge: Cambridge University Press, 2014.

Labeyrie, A., S. G. Lipson, and P. Nisenson. *An Introduction to Optical Stellar Interferometry.* Cambridge: Cambridge University Press, 2014.

Lawson, P. R. "Notes on the History of Stellar Interferometry: Principles of Long Baseline Stellar Interferometry." NASA-JPL Report: 00–009 (2000).

Lena, P. "Adaptive Optics: A Breakthrough in Astronomy." *Experimental Astronomy* 26, no. 1–3 (2009): 35–48.

Millour, F. "All You Ever Wanted to Know About Optical Long Baseline Stellar Interferometry, But Were Too Shy to Ask Your Adviser." *New Astronomy Reviews* 52, no. 2–5 (2008): 177–185.

Monnier, J. D. "Optical Interferometry in Astronomy." *Reports on Progress in Physics* 66 (2003): 789–857.

Saha, S. K. "Modern Optical Astronomy: Technology and Impact of Interferometry." *Reviews of Modern Physics* 74 (2002): 551–600.

Silva, I. and O. Freire. "The Concept of the Photon in Question: The Controversy Surrounding the HBT Effect circa 1956–1958." *Historical Studies in the Natural Sciences* 43 (2013): 453–491.

Quirrenbach, A. "The Development of Astronomical Interferometry." *Experimental Astronomy* 26 (2009): 49–63.

Tyson, R. K. *Principles of Adaptive Optics.* 2nd ed. Boston: Academic Press, 1998.

6. Across the Universe

Adhikari, R. X. "Gravitational Radiation Detection with Laser Interferometry." *Reviews of Modern Physics* 86, no. 1 (2014): 121–151.

Bracewell, R. N. "Detecting Non-solar Planets by Spinning Infrared Interferometer." *Nature* 274, no. 5673 (1978): 780–781.

Lacour, S. et al. "First Direct Detection of an Exoplanet by Optical Interferometry, Astrometry and K-band Spectroscopy of HR 8799 e." *Astronomy & Astrophysics* 623 (2019): L11.

Levin, J. *Black Hole Blues: And Other Songs from Outer Space.* First Anchor Books ed. New York: Anchor Books, 2017.

Mayor M. and D. Queloz. "A Jupiter-mass Companion to a Solar-type Star." *Nature* 378, no. 6555 (1995): 355–359.

Perryman, M. "The History of Exoplanet Detection." *Astrobiology* 12 (2012): 928–939.

Reitze, D. H., P. R. Saulson, and H. Grote. *Advanced Interferometric Gravitational-wave Detectors* (100 Years of General Relativity, vol. 5). Singapore: World Scientific Publishing Co., 2019: 2424–8223.

Sherrill, T. J. "Imaging Earthlike Exoplanets." *American Scientist* 93, no. 6 (2005): 516–523.

7. Two Faces of Microscopy

"Abbe, Ernst." *Complete Dictionary of Scientific Biography*, vol. 1. Charles Scribner's Sons, 2008: 6–9. *Gale eBooks.*

"Fraunhofer, Joseph." *Complete Dictionary of Scientific Biography*, vol. 5. Charles Scribner's Sons, 2008: 142–144. *Gale eBooks.*

Jackson, M. W. *Spectrum of Belief: Joseph von Fraunhofer and the Craft of Precision Optics (Transformations).* Cambridge, MA: MIT Press, 2000.

Kohler, H. "On Abbes Theory of Image-formation in the Microscope." *Optica Acta* 28, no. 12 (1981): 1691–1701.

Mertz, J. *Introduction to Optical Microscopy.* 2nd ed. Cambridge: Cambridge University Press, 2019.

Sweetnam, G. K. *The Command of Light: Rowland's School of Physics and the Spectrum* (Memoirs of the American Philosophical Society held at Philadelphia for Promoting Useful Knowledge: v. 238). Philadelphia: American Philosophical Society, 2000.

8. Holographic Dreams of Princess Leia

Johnston, S. F. *Holographic Visions: A History of New Science.* Oxford: Oxford University Press, 2006.

Johnston, S. F *Holograms: A Cultural History.* Oxford: Oxford University Press, 2016).

Leith, E. N. "Electro-optics and How it Grew in Ann Arbor." *Optical Spectra* (1971): 25–26.

Leith, E. N. "The Legacy of Dennis Gabor." *Optical Engineering* 19 (1980): 633–635.

Leith, E. N. "A Short History of the Optics Group of the Willow Run Laboratories." In *Trends in Optics*, edited by A. Consortini, 1–26., Cambridge MA: Academic Press, Elsevier, 1996.

Leith, E. N. "Overview of the Development of Holography." *The Journal of Imaging Science and Technology* 41 (1997): 201–204.

Leith E. N. and J. Upatnieks. "Wavefront Reconstruction with Diffused Illumination and Three Dimensional Objects." *Journal of the Optical Society of America* 54 (1964): 1295–1301.

Leith E. N. and J. Upatnieks. "Photography by Laser." *Scientific American* 224 (1965): 24–36.

Leith, E. N. and J. Upatnieks. "Holography at the Crossroads." *Optical Spectra* 4 (1970): 21.

Leith E. N. and J. Upatnieks. "Progress in Holography." *Physics Today* 25 (1972): 28–34.

9. Photon Interference

Bell, J. S., R. A. Bertlmann, and A. Zeilinger. *Quantum [Un]speakables: From Bell to Quantum Information.* Berlin: Springer, 2002.

Bernstein, J. *Quantum Profiles.* Princeton: Princeton University Press, 2001.

Bohm, D. and Y. Aharonov. "Discussion of Experimental Proof for the Paradox of Einstein, Rosen, and Podolsky." *Physical Review* 108, no. 4 (1957): 1070–1076.

Flamini, F., N. Spagnolo, and F. Sciarrino. "Photonic Quantum Information Processing: A Review." *Reports on Progress in Physics*, Review 82, no. 1 (2019): 016001.

Fox, M. *Quantum Optics: An Introduction.* Oxford: Oxford University Press, 2006.

Furusawa, A. and N. Takei, "Quantum Teleportation for Continuous Variables and Related Quantum Information Processing." *Physics Reports: Review Section of Physics Letters* 443, no. 3 (2007): 97–119.

Gea-Banacloche, J. "Optical Realizations of Quantum Teleportation." In *Progress in Optics, Vol. 46*, edited by E. Wolf, 311–353. Amsterdam: Elsevier, 2004.

Greenberger, D. M., M. A. Horne, and A. Zeilinger. "Multiparticle Interferometry and the Superposition Principle." *Physics Today* 46, no. 8 (1993): 22–29.

Horne, M., A. Shimony, and A. Zeilinger. "Quantum Optics – 2-particle Interferometry." *Nature* 347, no. 6292 (1990): 429–430.

Ou, Z.-Y. J. *Multi-photon Quantum Interference.* New York: Springer, 2007.

Ou, Z.-Y. J. *Quantum Optics for Experimentalists.* Singapore: World Scientific, 2017.

Pan, J.-W. Z.-B. Chen, C.-Y. Lu, H. Weinfurter, A. Zeilinger, and M. Zukowski. "Multiphoton Entanglement and Interferometry." *Reviews of Modern Physics* 84, no. 2 (2012): 777–838.

Pirandola, S., J. Eisert, C. Weedbrook, A. Furusawa, and S. L. Braunstein, "Advances in Quantum Teleportation." *Nature Photonics* 9, no. 10 (2015): 641–652.

Reid, M. D., P. D. Drummond, W. P. Bowen, E. G. Cavalcanti, P. K. Lam, H. A. Bachor, U. L. Andersen, and G. Leuchs. "Colloquium: The Einstein-Podolsky-Rosen Paradox: From Concepts to Applications." *Reviews of Modern Physics* 81, no. 4 (2009): 1727–1751.

Walls, D.F. and G. J. Milburn, eds. *Quantum Optics.* Berlin, Heidelberg: Springer, 2008.

Wick, D. *The Infamous Boundary: Seven Decades of Heresy in Quantum Physics.* New York: Springer, 1996.

Zeilinger, A. *Dance of the Photons: From Einstein to Quantum Teleportation.* Macmillan, New York: Farrar, Straus and Giroux, 2010.

10. The Quantum Advantage

Aaronson, S. and A. Arkhipov. "The Computational Complexity of Linear Optics." In *43rd ACM Symposium on Theory of Computing,* San Jose, CA, June 6–8, 2011, (Assoc. Computing Machinery, in Annual ACM Symposium on Theory of Computing, 2011): 333–342.

Barz, S. "Quantum Computing with Photons: Introduction to the Circuit Model, the One-way Quantum Computer, and the Fundamental Principles of Photonic Experiments." *Journal of Physics B: Atomic, Molecular, and Optical Physics* 48, no. 8 (2015): 083001.

Bogaerts, W., D. Perez, J. Capmany, D. A. B. Miller, J. Poon, D. Englund, F. Morichetti, and A. Melloni. "Programmable Photonic Circuits." *Nature,* Review 586, no. 7828 (2020): 207–216.

Briegel, H. J., D. E. Browne, W. Dur, R. Raussendorf, and M. Van den Nest. "Measurement-based Quantum Computation." *Nature Physics,* Review 5, no. 1 (2009): 19–26.

Brod, D. J., E. F. Galvao, A. Crespi, R. Osellame, N. Spagnolo, and F. Sciarrino. "Photonic Implementation of Boson Sampling: A Review." *Advanced Photonics* 1, no. 3 (2019): 034001.

Broome, M. A., A. Fedrizzi, S. Rahimi-Keshari, J. Dove, S. Aaronson, T. C. Ralph, and A. G. White. "Photonic Boson Sampling in a Tunable Circuit." *Science* 339, no. 6121 (2013): 794–798.

Deutsch, D. "Quantum Computational Networks." *Proceedings of the Royal Society of London Series A: Mathematical, Physical, and Engineering Sciences* 425, no. 1868 (1989): 73–90.

Knill, E., R. Laflamme, and G. J. Milburn. "A Scheme for Efficient Quantum Computation with Linear Optics." *Nature* 409, no. 6816 (2001): 46–52.

Kok, P., W. J. Munro, K. Nemoto, T. C. Ralph, J. P. Dowling, and G. J. Milburn. "Linear Optical Quantum Computing with Photonic Qubits." *Reviews of Modern Physics* 79, no. 1 (2007): 135–174.

Milburn, G. "Quantum Optical Fredkin Gate." *Phys. Rev. Lett.* 62 (1989): 2124.

Morton, O. "The Computable Cosmos of David Deutsch." *American Scholar* 69, no. 3 (2000): 51–67.

O'Brien, J. L. "Optical Quantum Computing." *Science* 318, no. 5856 (2007): 1567–1570.

O'Brien, J. L., G. J. Pryde, A. G. White, T. C. Ralph, and D. Branning. "Demonstration of an All-optical Quantum Controlled-NOT Gate." *Nature* 426, no. 6964 (2003): 264–267.

O'Brien, J. L., A. Furusawa, and J. Vuckovic. "Photonic Quantum Technologies." *Nature Photonics*, Review 3, no. 12 (2009): 687–695.

Ralph, T. C. and G. J. Pryde. "Optical Quantum Computation." In *Progress in Optics, Vol. 54*, edited by E. Wolf (2010): 209–269.

Reck, M., A. Zeilinger, H. J. Bernstein, and P. Bertani. "Experimental Realization of any Discrete Unitary Operator." *Physical Review Letters* 73, no. 1 (1994): 58–61.

Shor, P. W. "Scheme for Reducing Decoherence in Quantum Computer Memory." *Physical Review A* 52, no. 4 (1995): R2493–R2496.

Slussarenko, S. and G. J. Pryde. "Photonic Quantum Information Processing: A Concise Review." *Applied Physics Reviews*, Review 6, no. 4 (2019): 041303.

Spring, J. B., I. A. Walmsley, et al. "Boson Sampling on a Photonic Chip." *Science* 339, no. 6121 (2013): 798–801.

Tan, S.-H. and P. P. Rohde. "The Resurgence of the Linear Optics Quantum Interferometer − Recent Advances & Applications." *Reviews in Physics* 4 (2019): 100030.

Tanzilli, S., A. Martin, F. Kaiser, M. P. De Micheli, O. Alibart, and D. B. Ostrowsky. "On the Genesis and Evolution of Integrated Quantum Optics." *Laser & Photonics Reviews*, Review 6, no. 1 (2012): 115–143.

Wang, J. W., F. Sciarrino, A. Laing, and M. G. Thompson. "Integrated Photonic Quantum Technologies." *Nature Photonics* 14, no. 5 (2020): 273–284.

Zhong, H.-S., J.-W. Pan, et al. "Quantum Computational Advantage Using Photons." *Science* 370, no. 6523 (2020): 1460.

Index